VOLUME ONE HUNDRED AND ELEVEN

Advances in
COMPUTERS

Blockchain Technology: Platforms,
Tools and Use Cases

VOLUME ONE HUNDRED AND ELEVEN

ADVANCES IN
COMPUTERS

Blockchain Technology: Platforms, Tools and Use Cases

Edited by

PETHURU RAJ
Vice President
Site Reliability Engineering (SRE) Division,
Reliance Jio Infocomm. Ltd. (RJIL), Bangalore, India

GANESH CHANDRA DEKA
Deputy Director (Training), Directorate General of Training
Ministry of Skill Development & Entrepreneurship,
Government of India, New Delhi, India

ACADEMIC PRESS

An imprint of Elsevier

ISBN: 978-0-12-813852-6
ISSN: 0065-2458

For information on all Academic Press publications
visit our website at https://www.elsevier.com/books-and-journals

Working together
to grow libraries in
developing countries

www.elsevier.com • www.bookaid.org

Publisher: Zoe Kruze
Acquisition Editor: Zoe Kruze
Editorial Project Manager: Shellie Bryant
Production Project Manager: James Selvam
Cover Designer: Greg Harris

Typeset by SPi Global, India

CONTENTS

4. Applications of Blockchain in the Financial Sector and a Peer-to-Peer Global Barter Web 99

Kazuki Ikeda and Md-Nafiz Hamid

5. Blockchain for Business: Next-Generation Enterprise Artificial Intelligence Systems 121

Melanie Swan

PREFACE

Blockchain technology offers a wide variety of uses and the number of systems which employ it as an alternative controller of traffic has increased over the past decade. It attracts enormous international interest. Virtual currencies are widely used for multiple payments. Some countries have considered introducing the Blockchain technology to public services. Indeed, Estonia issues e-Residency, a digital ID that offers people all over the world to start up a location-independent business. The citizens can use it for various procedures including election, tax payment, bank transfer, and so forth. This edited book having eight chapters discusses about the various aspects of Blockchain technology.

Lots of organizations have already realized the potential of Blockchain Technology in Healthcare for securing the Healthcare data. Chapter 1 of this edited book presents some use cases of Blockchain technology in the healthcare sector. The authors have highlighted key design considerations for developing a Blockchain DApp (Decentralized Application) for Healthcare and describe the key lessons learned from their experience.

The researchers are putting their efforts to have a combined Automotive and Blockchain technology, but most of them considered applications, services, and smart contracts based on applications. In Chapter 2, the Local Dynamic Blockchain (LDB) and Main Blockchain (MB) with a secure and unique crypto ID, called Intelligent Vehicle Trust point (IVTP) to ensure trustworthiness among vehicles, are discussed.

Connecting the Blockchain to ERP across companies will lead to a new kind of technology solution that can be rightly called as Blockchain ERP or BERP. Chapter 3 discusses about the prospect of Blockchain technology in ERP.

The smart networks have made it possible for value transfer by the network itself without third-party intervention. Intelligence is built directly into the network's operations through a sophisticated protocol that automatically identifies, validates, confirms, and routes transactions within the network. Chapter 4 discusses about the various aspects of smart network.

Chapter 5 discusses about the "algorithmic trust" that significantly distinguishes itself from the more traditional typology of trust that was initially only between human agents.

Finally, Chapter 6 discusses how and under which conditions the use of Blockchain can be useful for a particular use case. General plausibility arguments are provided, including the details of use cases developed within prototypes as performed at the University of Zurich and others.

Chapter 7 discusses the security and privacy of Blockchain technology from both classical and quantum computing viewpoints. A novel full-quantum Blockchain model accommodating the quantum states in a peer-to-peer way is included in line with a growing sense of hope for the quantum era.

Chapter 8 presents some results of social experiments and evaluations of the proposed barter system. A brief review on several major Blockchain applications from viewpoints of both software and advanced use cases is presented.

PETHURU RAJ PH.D

GANESH CHANDRA DEKA

CHAPTER ONE

Blockchain Technology Use Cases in Healthcare

Peng Zhang*, Douglas C. Schmidt*, Jules White*, Gunther Lenz[†]
*Vanderbilt University, Nashville, TN, United States
[†]Varian Medical Systems, Palo Alto, CA, United States

Contents

Advances in Computers, Volume 111
ISSN 0065-2458
https://doi.org/10.1016/bs.adcom.2018.03.006

1

Abstract

Blockchain technology alleviates the reliance on a centralized authority to certify information integrity and ownership, as well as mediate transactions and exchange of digital assets, while enabling secure and pseudoanonymous transactions along with agreements directly between interacting parties. It possesses key properties, such as immutability, decentralization, and transparency, which potentially address pressing issues in healthcare, such as incomplete records at point of care and difficult access to patients' own health information. An efficient and effective healthcare system requires interoperability, which allows software apps and technology platforms to communicate securely and seamlessly, exchange data, and use the exchanged data across health organizations and app vendors. Unfortunately, healthcare today suffers from siloed and fragmented data, delayed communications, and disparate workflow tools caused by the lack of interoperability. Blockchain offers the opportunity to enable access to longitudinal, complete, and tamper-aware medical records that are stored in fragmented systems in a secure and pseudoanonymous fashion. This chapter focuses on the applicability of Blockchain technology in healthcare by (1) identifying potential Blockchain use cases in healthcare, (2) providing a case study that implements Blockchain technology, and (3) evaluating design considerations when applying this technology in healthcare.

1. INTRODUCTION

Blockchain is a platform that alleviates the reliance on a single, centralized authority, yet still supports secure and "trustless" transactions directly between interacting entities [1]. It offers decentralization, immutability, and consensus via cryptography and game theory. This technology provides the foundations for a number of application domains, including cryptocurrency and *Decentralized Apps* (DApps) [2].

Smart contracts are enhancements to Blockchain technologies, as implemented in the Ethereum Blockchain [3], that provide code to directly control the exchanges or redistributions of digital assets (such as cryptotokens or some pieces of data) between two or more parties according to certain rules or agreements previously established between involved participants. Smart contracts can store data objects and define operations on the data, enabling development of DApps to interact with Blockchains, and provide seamless services to the application users. In the domain of healthcare, smart contracts can be applied to create secure and effective technical infrastructures to enhance care coordination and quality and thus improve the wellbeing of individuals and communities [4]. Ideally, software apps and technology platforms in an interoperable healthcare environment should be able to communicate securely, exchange data, and use the exchanged data across health organizations and app vendors [5]. These health systems should also ensure effective care delivery for individuals and communities by allowing care providers to collaborate within and beyond organizational boundaries, e.g., by mediating secure access to electronic health records (EHRs).

Healthcare researchers and practitioners today, however, struggle with fragmented and siloed data, delayed communications, and disparate workflow tools. On the one hand, providers feel reluctant to exchange data due to (1) perceptions that patient health and identification information safe-keeping regulations prevent such sharing (even anonymized) and (2) potential liability and financial consequences associated with data sharing [6,7]. On the other hand, vendor-specific and incompatible health systems create gaps in healthcare communications, making it hard to coordinate and provide patient-centric care [8].

A key problem in production healthcare systems today is the lack of secure links that can connect all independent health systems together to establish an end-to-end reachable network [9] while protecting healthcare professionals with some level of anonymity (privacy). Although data standards like HL7 [10] and FHIR [11] provide basic interoperability for data exchange between trusted systems, this level of interoperability is limited to the implemented standards and requires data mapping between systems in most if not all cases. Maintainability of these systems is also hard to achieve since an interface change on one system requires other parties in the trusted network to adapt the change as well.

For example, Fig. 1 depicts barriers to achieving healthcare system interoperability, including incompatible software (such as vendor-locked EHR systems) and the lack of access to data outside a healthcare environment (such as a firewall protected clinic database or a patient-collected mobile health data). A promising solution to these problems involves the application of Blockchain technology, which provides "trustless" transactions via decentralization with pseudoanonymity. Complexities within the healthcare industry, however, yield additional challenges to employing Blockchain technology. This chapter explores the fundamental properties of Blockchain technology that can assist in establishing these trusted links and other capabilities in several potential use cases and analyzes the key design considerations for creating production DApps in the healthcare domain.

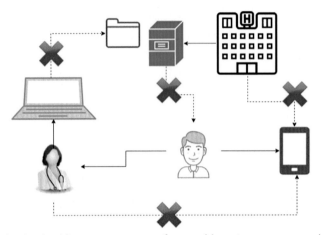

Fig. 1 Production healthcare systems are often unable to interoperate, and access to data records is disabled by disparate and siloed systems. © *[2017] IEEE. Reprinted with permission from P. Zhang, D.C. Schmidt, J. White, G. Lenz, M. Walker, Metrics for assessing Blockchain-based healthcare decentralized apps, Proceedings of the IEEE Healthcom 2017, October 12–15, 2017, Dalian, China.*

The remainder of this chapter is organized as follows: Section 2 identifies pressing issues in healthcare, focusing on interoperability and patient-centered care. Section 3 then describes seven specific healthcare scenarios, or use cases, where Blockchain technology can be leveraged to alleviate some of the major challenges associated. Section 4 explores four healthcare-specific challenges faced by Blockchain-based systems and their design implications. Section 5 presents a case study prototype we developed as an example that addresses the challenges. Section 6 uses the case study to highlight key design considerations for developing Blockchain-based DApps for healthcare. Section 7 summarizes key lessons learned from our experience. Section 8 provides an outlook on future research directions of Blockchain technology in the healthcare space.

2. PRESSING ISSUES IN HEALTHCARE

This section explores pressing issues in healthcare today, focusing on interoperability challenges that limit data sharing and impede patient-centered care fostered by healthcare interoperability that would otherwise allow patients to access and control their own health information.

2.1 Interoperability

Healthcare interoperability describes the ability for heterogeneous information technology systems and software applications, such as the EHR system, to communicate, exchange data, and use the exchanged data [12]. Allowing information systems to work together within and across organizational boundaries are paramount to improve effective care delivery for individuals and communities [13]. For example, interoperability enables providers to securely and scalably share patient medical records with one another (given patient permissions to do so), regardless of provider location and trust relationships between them.

Secure and scalable data sharing is essential to provide effective collaborative treatment and care decisions for patients. Data sharing helps improve diagnostic accuracy [14] by gathering confirmations or recommendations from a group of medical experts, as well as preventing inadequacies [15] and errors in treatment plan and medication [16,17]. Likewise, aggregated intelligence and insights [18–20] help clinicians understand patient needs and in turn apply more effective treatments. For example, groups of physicians with different specialties in cancer care form tumor boards that meet regularly to discuss cancer cases, share knowledge, and determine effective cancer treatment and care plans for patients [21]. As another example, if a cancer patient under treatment is admitted to the emergency room (ER) in a different hospital, then it would be critical for the ER provider to access the patient's medical records to identify potential drug interactions; while the patient's primary cancer care provider should be notified that the patient is being treated in the ER.

Despite the importance of medical data sharing, today's healthcare systems frequently require patients to obtain and share their own medical records with other providers either via physical paper copies or electronic hard disk copies. This process of obtaining and sharing medical records is ineffective for the following reasons:

- *It is slow* since copies of medical data must be prepared, delivered, and picked up by patients. The law allows providers up to 30 days to supply medical records to patients, although some providers may only take 5–10 business days to prepare noncritical health records [22].
- *It is insecure* because data copies may be lost or stolen during their physical transmission by patients from one location to another.
- *It is incomplete* since as patient health history may be fragmented because their data are stored in disparate and siloed systems. There is no single source that stores all the medical records of an individual, so patients must

therefore be responsible for keeping track of when and where they received health services in order to request copies of their medical history.

- *It lacks context* since today's healthcare systems are provider centric instead of patient centric, thereby preventing patients from taking control of their own health records and having knowledge of what is done to their data or who has accessed their data [23].

The ineffective data sharing process in healthcare results in part from the lack of trust between providers and the lack of interoperability between health IT systems and applications today. Healthcare interoperability comprises of three levels: foundational, structural, and semantic, ordered the lowest to the highest fidelity [24], as summarized in Table 1.

Foundational interoperability enables data exchanges between healthcare systems. It does not require providers receiving the data to be able to interpret the data. *Structural interoperability* additionally defines formats for exchanged clinical data and ensures that received data are preserved and can be interpretable at the data field level using predefined formats. *Semantic interoperability* demands the interpretability of exchanged data by not only syntax (structure) but also semantics (meaning) of the data.

These three levels help ensure that disparate health systems and applications deliver information with requisite data quality and safety. Foundational and structural interoperability are prerequisites for semantic interoperability, which is hardest to achieve but most desired to advance quality of care. This chapter focuses on exploring Blockchain-based use cases that address foundational and structural interoperability from the technical infrastructure's perspective. Semantic interoperability requires clinical domain knowledge and medical policies that enforce the adoption of ontologies, i.e., common data standards, to interpret myriad sources of health information, which is beyond the scope of this chapter.

Table 1 Summary of the Three Levels of Interoperability

Interoperability Level	Summary
Foundational	Data exchange is enabled; does not require data interpretation
Structural	Defines formats for the exchanged data
Semantic	Requires interpretation of the exchanged data

2.2 Patient-Centered Care

The healthcare industry is shifting from volume-based care (fee-for-service), in which providers are incentivized to provide more treatments because payment is proportional to the quantity of care, to value-based care that promotes patient-centered care with higher quality (hence "value"), in which patients are informed and involved in clinical decision making [25]. In patient-centered care model, patients are capable of incorporating patient reported experience measures and patient recorded outcome measures [26], such as symptoms or health status, collected from their wearable and mobile devices into their medical history. Patients should also be given easy access to their medical records with a comprehensive view of their entire health history, which could potentially reduce information fragmentation and inaccuracy caused by communication delays or coordination errors and in turn improve the care continuity and quality [27].

Ideally, all health systems (regardless of the type of care settings) would provide automatic notifications for patients to access their clinical data in (near) real time, e.g., as records are entered into the system or when lab results are available. In addition, in patient-centric care it is necessary for patients to control when and to whom their health data are shared and to choose what pieces of information they would be willing to share. Healthcare systems today, however, do not provide the means for patients to modify or revoke a provider's access to their data.

As a result, once a provider has cared for a patient or has obtained access to patient data that data are permanently in possession of the provider. When a patient visits different providers many times throughout their lifetime, their health and other sensitive personal data are available at several sites. This diffusion increases the risk of data theft because it only takes one provider lacking sufficient and up-to-date security practices to put patient information vulnerable to attack (with the assumption that no malpractice or unethical usage of patient data are involved, of course). Alternately, a patient may wish to release their medical records to a new provider, which is not easily accomplished today as discussed in Section 2.1.

Interoperability is also fundamental to support a patient-centric model that improves quality of care for individual patients. In practice, barriers exist in the healthcare technical infrastructure that impedes interoperability and thus patient-centered care, including (but not limited to):

- *Information security and privacy concerns.* Despite the need for data sharing, it increases the risk of sensitive data breach without a highly secure infrastructure in place [7]. Providers could face severe financial and legal consequences [6] when data are compromised.

- *Lack of trust between providers.* Because of security regulations, care providers must be able to identify other providers and also trust their identities before any patient health-related communication occurs [28]. Trust relationships often exist between in-network providers and/or health organizations but they are particularly difficult to establish when the data receiving care office do not use the same health system with a shared provider directory, such as in a private practice or a hospital network.
- *Scalability concerns.* Medical data may contain large volumes of data like medical images, especially in cancer patients or patients with chronic conditions. These large-scale datasets are difficult to share electronically due to limitations in bandwidth or restrictive firewall settings, such as in rural areas [29].

Below we first present six potential use cases of Blockchains in the healthcare space and then focus on a concrete case study that demonstrates what design decisions can be made to leverage Blockchain-related technologies to address these interoperability challenges.

3. BLOCKCHAIN USE CASES IN HEALTHCARE

This section explores seven Blockchain use cases focusing on various concerns in healthcare as summarized in Table 2.

Table 2 A Summary of Seven Healthcare Use Cases That Blockchain Technology Can Address

Sections	Use Case Summary
3.1	Prescription tracking to detect opioid overdose and overprescription
3.2	Data sharing to incorporate telemedicine with traditional care
3.3	Sharing cancer data with providers using patient-authorized access
3.4	Cancer registry sharing to aggregate observations in cancer cases
3.5	Patient digital identity management for better patient record matching
3.6	Personal health records for accessing and controlling complete health history
3.7	Health insurance claim adjudication automation to surface error and fraud

3.1 Opioid Prescription Tracking

It is widely known that there is an opioid epidemic present in the United States [30]. While many efforts are being made to address this crisis (e.g., the Drug Supply Chain Safety Act (DSCSA) [31], the President's Commission on Combating Drug Addiction and the Opioid Crisis [32], and numerous prescription awareness campaigns [33,34]), our current prescription tracking system still lacks the technology to do so effectively. Data hoarding, doctor shopping [35], provider ignorance, vulnerable and centralized data, and overprescription riddle the current prescription opioid marketplace. The decentralization and auditability of Blockchain technology provide a promising approach to prescription monitoring that not only makes prescriptions safer, but also provides incentives for writing fewer prescriptions.

Healthcare providers today are incentivized to prescribe opioids to patients. For example, providers incur less face time with patients, fewer costs associated with patient care, and thus greater profits from higher returns. Likewise, pharmacies are incentivized to produce and distribute opioids since the more they sell, the higher their bottom line, and the greater their return to shareholders. Moreover, patients are incentivized to consume opioids. In the treatment of pain, physical therapy or postsurgery recovery can be frustrating and riddled with disappointment [36]. Opioids provide a short-term relief, at the cost of addicting a patient. This self-fulfilling cycle can thus benefit from a technological solution that realigns these incentives.

To offset these incentives contributing to the rise of the opioid epidemic, a Blockchain-based system can establish a trusted network of hospitals and pharmacies to store opioid-associated transactions (including prescriptions, fulfillment, etc.) in a secure and accountable manner. Such a distributed and shared permissioned Blockchain platform allows for loosely coupled providers to access other data silos without explicit trust relationships between each other. Stakeholders within the system (hospitals, pharmacies, etc.) are likewise incentivized to onboard new members to the consortium because with each additional member, they can form a more complete dataset. Rules can be mutually predefined so the consortium can securely onboard new provider members to the system.

By distributing knowledge that an opioid transaction has occurred, rather than the entirety of the specific content of that transaction, this type of ecosystem can remedy a number of the problems in the current opioid system. More complete opioid prescription history can be available to detect overly prescribed opioid by providers and also patterns of doctor shopping in patients. Consequently, providers will be incentivized to meet the requirements to join the consortium of providers through potential access to data that will increase the quality of their care. Most importantly, by tracking the history of opioid prescriptions, patients will receive care more appropriate to their condition and thus be steered away from the dangers of opioids toward less addictive and thus longer lasting treatment actions.

3.2 Data Sharing Between Telemedicine and Traditional Care

Traditionally, telemedicine offers widely accessible care to patients who are located in remote areas far away from local health facilities or in areas of with shortages of medical staff. Today, it has becoming increasingly popular among patients who wish to receive convenient medical care [37]. Connected patients can avoid wasting time waiting at a doctor's office and get immediate treatment for minor but urgent conditions on demand [38]. Owing to the growing accessibility to smart mobile and telemedicine devices, many companies offer 24/7 continuous access to care, and many user-friendly apps have been created for patients to monitor, manage, and report their health using technology [39]. For example, Apple Health [40] app allows patients to connect to equipment for measuring vitals and store these data on their iPhones. These records can then be reported to the provider as needed.

Telemedicine services are usually equipped with more advanced technologies and are much more far-reaching compared to traditional physical health services. Owing to these on-demand services, it is common for providers from different regions or networks to treat patients, resulting in reduced care continuity. Health data collected during telemedicine care episodes may be inaccessible by patients' primary care providers, which creates an incomplete medical history and in turn risks the overall quality of care [41].

By removing the need for a third-party authority and empowering direct interactions between involved participants, Blockchain technology can potentially bridge the communication barrier between these providers. Blockchain technology alone, however, cannot address the complex data sharing challenge—it must be incorporated into existing disparate health systems and clinical data standards. Fig. 2 shows a high-level conceptual infrastructure where a Blockchain (represented as the dashed ellipse) is connected to disparate health database systems (represented as cylinder database objects).

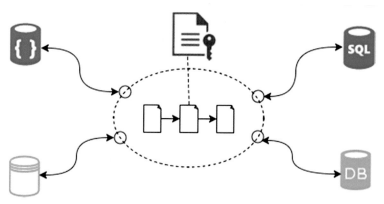

Fig. 2 A high-level blockchain-based conceptual infrastructure to connect disparate health database systems and record data exchange history.

Each database system is shown in Fig. 2 opens up a new secure data channel (as represented by small circles on the ellipse border), similar to what is used to share data with other akin systems. A smart contract (the keyed file symbol) is then used to govern the data transactions between these systems based on mutual agreements and also create an immutable history of all the transaction records. In practice, a robust architecture will include many more design components than what is shown in Fig. 2.

3.3 Patient-Controlled Cancer Data Sharing

Cancer diagnoses and treatment plans are rarely black-and-white, i.e., they involve many considerations due to the complexity of tumor cases and the number of available treatment options [42]. Getting fresh perspectives from different specialists can help narrow down the options and may shorten the time from (suspected) diagnosis to treatment for a cancer patient. In the United States today, most hospitals have included at least one tumor board, which is a multidisciplinary team of medical, surgical, radiation oncologists, and other specialists and care providers, to review and discuss individual cancer patient's condition and treatment options in depth [2]. Despite the increasing effort to encourage cancer care collaboration among oncologists, patients, and families remain passive in the decision-making process. A large-scale enterprise hospital may involve specialists from a wider range of disciplines, whereas a smaller-scale care center may have limited resources to expand their tumor boards. The quality of life that is important to cancer patients may be neglected due to patient disengagement.

In reality, patients may wish to reach out to a new provider for a second opinion on their medical conditions and/or treatment plan. To share critical data today, patients have to obtain copies of their medical reports from their current provider, which may include their family history, visit history, prescriptions, current diagnoses and treatment options, and so on. All reports will then be delivered physically to the new provider. In this technologically advanced society, patients with critical conditions should be least involved in the manual data sharing process and have the critical data shared in a timely manner to prevent delays in treatment. A patient-controlled data sharing feature is missing from the existing health systems for cancer patients to promptly request a second opinion and also selectively share information.

Instead of creating a new trusted "middle man" that mediates the establishment of trust relationships between providers/hospitals, Blockchain technology offers the opportunity for trustless exchange and disintermediation that allow existing trust relationships to be aggregated and propagated across various organizations and providers. This approach is similar to the patient referral process, except the referrals are not limited to a single provider's network. Instead, they could be expanded across different regions, states, and even countries. A Blockchain-based system can also capture existing trust relationships between patients and providers, allowing patients to decide which provider(s) can share their data.

3.4 Cancer Registry Sharing

Data sharing is especially critical in cancer care where cases are usually complex and cures are rarely one-size-fits-all [43]. Being able to share cancer data helps ensure the integrity of results obtained from clinical trials by enabling individual confirmation and validation, but it can also agglomerate intelligence gathered to reduce unwarranted repetition in clinical trials [18–20]. It allows distributed clinical trials to achieve a significant cohort size and thus speeds up the discovery of more effective cancer treatments. In the United States, only about 3% [44] of cancer patients are undergoing clinical trials today. As a result, most cancer patients receive treatments based on observations drawn from this small population of highly selective patients with different demographics, family medical history, secondary diagnoses, etc.

Population-based cancer registries are attempts to capture very rudimentary data from cancer incidences across geographic areas and for planning population-wide cancer control [45]. As with EHRs, cancer registries are often siloed and fragmented, which can similarly leverage Blockchain technology for expedited information exchange. In addition, with increased availability of richer data collected from many patients, artificial intelligence can be used to construct prognostic and predictive models for assisting care providers with decision support. A learning ecosystem can be designed using Blockchain technologies to share predictive models and collaboratively improve accuracies of learned medical insights.

3.5 Patient Digital Identity

A fundamental component in health information exchange is patient identification [46] matching, which finds a patient in a healthcare database using a unique set of data. Systems like the master patient index and enterprise patient master index [47] have been created to manage patient identities within a healthcare organization or within a trusted network. Despite the increased development effort, accurately and consistently matching patient data remain hard. Patient identity mismatching has contributed to duplicated patient records and incomplete or incorrect medical data [48].

One study [49] estimated that 195,000 deaths occur each year due to medical errors, with 10 of 17 errors being identity or "wrong patient errors." There are also significant costs to healthcare organizations who maintain these duplicate records and correct mistakenly merged errors [50] and also patients who experience repeated tests or treatment delays. In addition, these errors also impact reimbursement as claims may be denied due to "out of date or incorrect information" [51], not to mention the security risks involved when patients disclose their personal information.

Without common standards for collecting patient identifying information, even the same patient's identity can vary from one care facility to another. For instance, demographics data, such as name, date of birth, address, and Social Security number (SSN) are often used to register a patient [46]. However, names may be stored in various formats, such as legal first and last name, nickname and last name, with or without middle initial, and patients may share identical or similar names; similarly, date of birth can be entered into the system in multiple ways; address can change as patients move to a new location; and patients may refuse to provide their SSN or do not have an SSN. Furthermore, patient information manually entered into the system may contain typos or errors, and the more data collected, the more opportunities for mistakes [51]. Although within each organization patient demographics data may be collapsed into a single unique ID, the ID generally does not translate across organizations.

Without a functional, unified identity management system, patient identification schemes employed at various care sites may continuously experience incompatibility and patient matching problems, unless a patient exclusively receives care within one organization. In fact, the very nature of Blockchain incorporates such a decentralized, unified identity system. Many existing Blockchains use cryptographically secured addresses to represent identities. Each address is mathematically linked with a unique key that is used to easily verify the ownership of an address or an identity yet does not reveal any personal information relating to the individual. The decentralized and auditable characteristics of Blockchain can help enforce standardized verifiable identities for patients via a universal patient index registry sharable across all healthcare facilitates within a nation and beyond [52]. In case of lost or stolen keys, new addresses are also trivial to generate and reassign to patients.

3.6 Personal Health Records (PHRs)

Unlike the current standard practice of using provider-centric EHRs to maintain and manage patient data, PHRs are applications used by patients, the true data owners, to access and manage their health information. The ultimate goal for PHRs is to help patients securely and conveniently collect, track, and control their complete health records compiled from various sources, including provider visit data, immunization history, prescriptions records, physical activity data collected from Smartphone devices, and many more. PHRs enable patients to control how their health information is used and shared, verify the accuracy of their health records, and correct potential errors in the data [53]. Enterprises and technology companies, such as Apple and Microsoft, have begun exploring centralized solutions with their Apple Health [40] and Microsoft HealthVault [54] products. Centralized approaches do not resolve the data sharing problem at its core, however, and may therefore face similar hurdles as existing disparate EHR systems.

Blockchains, in contrast, allow distribution of control to individuals via decentralization enabled by consensus algorithms. By creating a widely accessible and secure data distribution service that connects to existing health systems, patients can easily aggregate their medical history without requesting a copy from every provider they have visited. Connections to personal smart devices are also possible as Blockchains remove the "distrust" between healthcare professionals and third-party health tracking apps and services. Furthermore, permission-based data distribution can be set up with smart contracts to guarantee that patients (1) remain in control of their data access, (2) are aware of the origin of aggregated data sources, and (3) are informed when their data are accessed by providers. Data origin and access history are made transparent to the patients through immutable audit logs to always keep patients up-to-date of when and by whom their health information is retrieved.

3.7 Health Insurance Claim Adjudication

Health insurance is used to protect individual assets from the devastating costs of a major accident, medical emergency, or treating a chronic disease and to ensure care is provided when needed. It can cover the cost of doctor visits, medical, and surgical expenses, depending on the type of health insurance coverage [55]. Patients may a portion of out-of-pocket costs at the point of care, but the remaining costs are submitted as claims to the health insurance companies. Providers are then reimbursed through the "claim adjudication" process, whereby the insurer determines their financial responsibility for the payment and the payment amount (if applied) directly made to the provider [56].

The insurer may decide to pay the claim in full, deny the claim, or reduce the amount paid to the provider. The decision to reduce a payment to the provider is typically made when the insurance company has determined that the billed service is inappropriate or medically unnecessary for the diagnosis or procedure codes. Therefore, "it is important to ensure that all claims submitted for payment are coded accurately. As soon as an insurance company receives a medical claim, they begin a thorough review. Sometimes even small errors such as a misspelled patient name may cause a claim to be rejected" [57]. The majority of claims today are processed automatically without manual intervention, but as claims become more complex and are plagued by error and fraud [58], claim adjudication remains a challenging process.

Currently, 22% of claims get rejected either because they are not received by the insurer or they contain defects, such as incomplete or incorrect demographic data or lack of proof supporting the services billed [59]. Fortunately, smart contracts present an opportunity for automating the adjudication process further by distributing and thus making claims transparent to the provider and insurer, exposing potential errors, and frauds that can be corrected or investigated in a much timelier manner. Another benefit of creating these preestablished agreements via smart contracts is to ensure involved participants are up-to-date and properly notified as policies or rules change.

4. HEALTHCARE INOPERABILITY CHALLENGES FACED BY BLOCKCHAIN-BASED APPLICATIONS

While it is important to understand the fundamental properties that Blockchains possess, it is also crucial to analyze domain-specific challenges that this technology may face to ensure the practicality of Blockchain use cases. This section examines four key interoperability challenges faced by Blockchain-based healthcare applications: system evolvability, data storage on Blockchain, healthcare information privacy, and system scalability, as summarized in Table 3.

4.1 Evolvability Challenge: Supporting System Evolution While Minimizing Integration Complexity

Many apps are created with the assumption that data are easy to change. This assumption is valid in most centralized systems where the data are managed by some trusted party, such as Amazon Web Services. In a Blockchain-based app with decentralized storage, however, data are difficult to modify en masse and its change history is recorded as an immutable log. A critical consideration when using Blockchain technology in health system design is thus to ensure that the data structures defined in a Blockchain (i.e., via smart contracts) are designed to facilitate evolution where needed to minimize change.

Table 3 Summary of the Interoperability Challenges Faced by Blockchain Technologies

Challenge	Section(s)	Challenge Summary
Evolvability	4.1 and 6.1	Blockchain is immutable but needs to support health system evolution
Storage	4.2 and 6.2	Inefficient and costly on-chain storage should be minimized yet still provide data access
Privacy	4.3 and 6.3	Data transparency of the Blockchain should be balanced with health privacy concerns
Scalability	4.4 and 6.4	Relevant data should be filtered out from all events recorded on the Blockchain

Although evolution must be supported, healthcare data must often be accessible from a variety of deployed systems that cannot easily be changed over time. Necessary data structures should therefore be designed in a way that are loosely coupled and minimize the usability impact of evolution on the clients that interact with data in the Blockchain. Section 6.1 shows how we design with the Abstract Factory pattern to facilitate evolution while minimizing the impact on dependent clients, focusing on a concrete evolution challenge involving entity creation and management of healthcare participants.

4.2 Storage Challenge: Minimizing Data Storage Requirements on the Blockchain

Healthcare applications typically serve thousands or millions of participants, including providers, patients, billing agents, and so on. It may incur enormous overhead when large volumes of data are being stored in a Blockchain—particularly if data normalization and denormalization techniques are not carefully considered. Not only is it costly to store these data, but also data access operations may fail if/when the cost exceeds Blockchain network defined data size limit. For instance, the Ethereum public Blockchain defines a "gas limit" that limits the capacity of data operations to prevent attacks manifest through infinite looping [3].

An important design consideration for Blockchain-based healthcare apps is thus to minimize data storage requirements on-chain yet provide sufficient flexibility to manage individuals' health concerns. Section 6.2 shows how to design smart contracts with the Flyweight pattern to ensure unique entity account creation that maximizes sharing of common intrinsic data across entities while still allowing extrinsic data to vary in individual entities.

4.3 Privacy Challenge: Balancing Data Sharing Capability With Privacy Concerns

The US Office of the National Coordination for Health Information Technology (ONC) has outlined a number of basic technical requirements for achieving interoperability [5]. These requirements include identifiability and authentication of all participants, ubiquitous and secure infrastructure to store and exchange data, authorization and access control of data sources, and the ability to handle data sources of various structures. Blockchains are emerging as promising and cost-effective means to meet some of these requirements due to their inherent design principles built upon secure cryptography and resilient peer-to-peer networks. Smart contract-enabled

Blockchain-based DApps provide the healthcare domain with capabilities like digital asset sharing (such as health records sharing) and audit trails of transaction history (such as data access records, which are essential for improving healthcare interoperability.

Although storing patient health data in the Blockchain may provide substantial potential benefits to interoperability and immediate availability, there are also significant risks due to the transparency of the data because each Blockchain manager/miner maintains a complete copy of Blockchain data. In particular, even when encryption is applied, it is still possible that the current encryption techniques may be broken in the future or that vulnerabilities in the encryption implementations used may lead to private information potentially being decryptable and compromised in the future. In Section 6.3, we discuss how to a Blockchain-based app using the Proxy pattern can facilitate interoperability while keeping sensitive patient data from being directly encoded in the Blockchain.

4.4 Scalability Challenge: Tracking Relevant Health Changes Scalably Across Large Patient Populations

Communication gaps and information sharing challenges are serious impediments to healthcare innovation and the quality of patient care. Providers, hospitals, insurance companies, and even departments within the same health organizations experience disconnectedness caused by delays or the lack of information flow. Patients are commonly cared for at various sources, such as private clinics, regional urgent care centers, enterprise hospitals, and telemedicine practice. A provider may serve hundreds or more patients whose associated health activities must be tracked. Without any activity monitoring or filtering mechanism implemented, it would require tremendous computational effort for a provider to review a patient's health transactions on-demand. Section 6.4 shows how a Blockchain-based app design using the Publisher–Subscriber pattern can be aid in scalably detecting and tracking relevant health changes.

5. CASE STUDY DAPP OVERVIEW

This section presents the structure and functionality of a case study DApp for Smart Health (DASH)[a] we developed to explore the efficacy of applying Blockchain technology to the healthcare domain. This prototype

---

[a] There is no relationship between our DASH app and the "Dash" cryptocurrency, even though they are both based on Blockchain technologies.

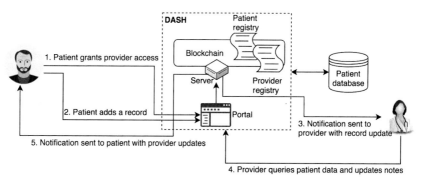

Fig. 3 DASH architecture overview. *Reprinted with permission from P. Zhang, J. White, D.C. Schmidt, G. Lenz, Applying software patterns to address interoperability in Blockchain-based healthcare apps, The 24th Pattern Languages of Programming Conference, October 22–25, 2017, Vancouver, Canada.*

was implemented on an Ethereum test Blockchain to emulate a minimal version of a personal EHR system. It provides a web-based portal for patients to self-report and access their medical records, as well as submit prescription requests. DASH also includes a staff portal for providers to review patient data and fulfill prescription requests based on permissions given by patients. Fig. 3 shows the structure and workflow of DASH.

The core user features supported in DASH can be summarized as the follows:

1. Patients can grant a provider permissions to access their health records or prescription requests via the DASH portal.
2. Patients can add a health record through a standardized, preformatted form through the DASH portal.
3. Health–related activities (i.e., prescription requests and health record additions) related to a patient are sent to providers with authorized access to the patient's data with secure notification messages.
4. Provider with authorized access to a patient's records can query, make changes, and upload physician notes to the data, as well as fulfill the patient's prescription requests.
5. Patients will be notified of any update to their health record performed by the provider.

DASH uses a *Patient Registry* contract to store a mapping, or relationship that links unique patient identifiers to their associated *Patient Account* contract addresses (locations). Each *Patient Account* contract maintains a list of healthcare providers (via unique provider identifiers) who are granted read/write access to the patient's medical records. At its current state, DASH is limited to only provide data access services to two types of users: patients and providers.[b] Patient health records are stored off-chain in a secure database, implementing the FHIR data standards. The reason for storing patient data in a centralized database is to simulate a data silo, as it is in today's health systems, to later on exercise data integration with other siloed databases. Our database server creates a secure socket to exchange permission-based tokenized access to patient data using standard public key cryptography. Provider and patient users who are members of DASH are each equipped with two secure cryptographic key pair for (1) encrypting and decrypting data references for authorizing access to a patient dataset and (2) signing new health records and verifying signatures to prevent tampering to the data.

6. DESIGN CONSIDERATIONS OF BLOCKCHAIN-BASED APPS FOR HEALTHCARE USE CASES

This section details the applications of familiar software patterns to address each healthcare interoperability challenge facing Blockchain apps. Namely, we focus on Abstract Factory, Flyweight, Proxy, and Publisher-Subscriber [60,61] patterns and demonstrate how to incorporate these patterns in the DASH design. Fig. 4 shows how an anatomy of DASH design with software pattern applications to address the following design challenges:

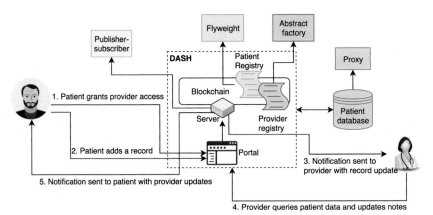

Fig. 4 Incorporating software patterns to DASH design. *Reprinted with permission from P. Zhang, J. White, D.C. Schmidt, G. Lenz, Applying software patterns to address interoperability in Blockchain-based healthcare apps, The 24th Pattern Languages of Programming Conference, October 22–25, 2017, Vancouver, Canada.*

---

[b] Naturally, there are other patterns relevant in this domain, which can be explored in future research.

- Abstract Factory assists with organization/individual account creation and management based on user types, especially in a structured or evolving organization.
- Flyweight ensures unique account creation on the Blockchain and maximizes sharing of common, intrinsic data.
- Proxy protects health information privacy while facilitating seamless interactions between separate components in the system to ensure appropriate levels of data accessibility.
- Publisher-Subscriber aids in scalably managing health change events and actively notifying healthcare participants when and only when relevant changes occur.

Detailed applications of these patterns are discussed in depth in the following subsections.

6.1 Evolvability: Maintaining Evolvability While Minimizing Integration Complexity

6.1.1 Design Problems Faced by Blockchain-Based Apps

As part of the initial user on boarding process, account creation, and management may occur on the Blockchain in order to define unique identifications for an organization or individual. Many organizations in the healthcare industry, such as enterprise hospitals and insurance companies, are hierarchical and evolving in nature. One design problem is thus to create an account structure that supports organization evolvability while minimizing integration complexity when new entities of different functions (e.g., a new division or department) are introduced. Specifically, the immutability property of Blockchain technology ensures that smart contract interface (including data member definitions and functions) cannot be modified. Each change to a smart contract must be deployed as a new contract object on the Blockchain and distributed among all the network nodes so it can be executed on-demand.

To minimize interface changes over time, it is important to create a modular design. As a concrete example, suppose one smart contract defines a function that manages interactions between different departments in a highly structured hospital. Without a modular design, many decisions must be made to identify the appropriate departmental accounts involved, thereby creating a large number of branching statements. As new departments are introduced, the decision-making code will likely have to change more than once, making each previous version of smart contract obsolete. A desired model should minimize interface changes to a contract.

6.1.2 Solution: Apply the Abstract Factory Pattern to Support Design Evolvability

Creating complex account structures in smart contracts for an evolving, hierarchical environment can be modularized by applying the Abstract Factory pattern [60]. This pattern allows DApps like DASH to delegate the responsibility for providing account creation services to an abstract "factory" object (which is a contract instance itself). A concrete factory object can then inherit methods from the abstract factory and customize them to create accounts for a specific set of related or interacting subentities. For instance, with this pattern implemented in the account structure, when a new department is introduced, the app simply creates another concrete factory object for the new entity without affecting existing accounts in other smart contracts. Fig. 5 shows the basic structure of this pattern in the context of DASH.

DASH's Blockchain component uses an *AbstractAccountFactory*, the abstract factory contract object, to define some common logic (i.e., *createAccount* and *createOrganization* methods) shared by all concrete factories (e.g., *ProviderFactory*, *InsuranceFactory*). When *Client*, i.e., DASH user,

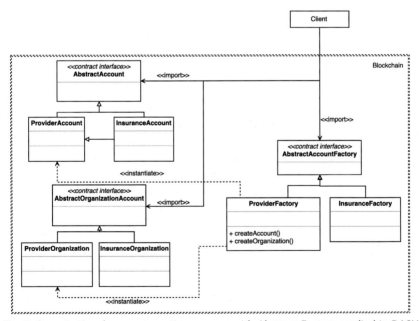

Fig. 5 Organizational account creation process with Abstract Factory applied in DASH design. *Reprinted with permission from P. Zhang, J. White, D.C. Schmidt, G. Lenz, Applying software patterns to address interoperability in Blockchain-based healthcare apps, The 24th Pattern Languages of Programming Conference, October 22–25, 2017, Vancouver, Canada.*

requests a new entity account structure to be created, regardless of the entity type, DASH server creates a concrete entity factory inherited from *AbstractAccountFactory* to create one account type for storing individual users and another account type for storing information regarding the entity organization. This corresponds to the instantiation of a *Provider* account and a *ProviderOrganization* for the *Provider* entity in the example from Fig. 5. The process of creating an account structure for an entity in this design is completely decoupled from already defined structures, thus leaving their corresponding account contract objects intact.

6.1.3 Consequences of Applying Abstract Factory

Without using a high-level abstraction, such as an abstract factory, creating constituent accounts for an entity would imply many if–else decisions made at runtime to determine the appropriate concrete account factory to execute. This tight coupling is cumbersome since *Client* (who interacts with the Blockchain component in the above example) has to be familiar with the implementation details in order to execute each concrete factory's methods properly. For example, to create an account structure for either *Provider* or *Insurance* entity in this case, *Client* must be exposed to both *ProviderFactory* and *InsuranceFactory* contracts and make decisions at runtime regarding which factory methods to call, as shown in Fig. 6. The example only presents two concrete entity examples, but as the number of entities scales up or as the departments within an organization scale out, the *Client* will have to keep track of an overwhelming amount of detailed implementation.

Entity creation with Abstract Factory introduces a loose coupling between the client (e.g., DApp server) and specific smart contract implementations (i.e., the Blockchain component). In particular, newly defined entity interactions can inherit from the abstract contract to preserve common properties and have entity-specific customizations. Because of the loose coupling, existing contracts remain unmodified, minimizing contract deprecation. The downside of using this design in a Blockchain, however, would be the extra storage (and therefore extra cost if used in a public Blockchain with a mining incentive) overhead incurred by an added layer of indirection for the abstract factory and its instantiation.

6.2 Storage: Storage Challenge: Minimizing Data Storage Requirements

6.2.1 Design Problem Faced by Blockchain-Based Apps

It may be inevitable that a Blockchain-based healthcare app requires some data to be maintained on-chain. However, a good design should minimize

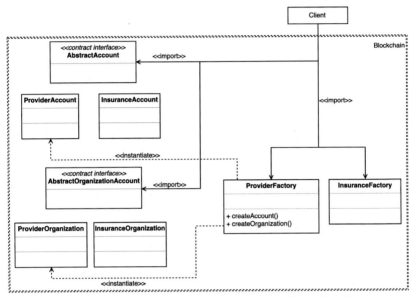

Fig. 6 Organizational account creation process without applying the Abstract Factory pattern. *Reprinted with permission from P. Zhang, J. White, D.C. Schmidt, G. Lenz, Applying software patterns to address interoperability in Blockchain-based healthcare apps, The 24th Pattern Languages of Programming Conference, October 22–25, 2017, Vancouver, Canada.*

the storage requirements by maximizing data sharing on the Blockchain to avoid storage and cost overhead. The price for Blockchain to have transparency and immutability is that all data and transaction records maintained in the Blockchain are replicated and distributed to every node in the network. To avoid unnecessary data storage, such as duplicated data or unattended data, it is therefore important to create a design that maximizes shared data on-chain.

If a Blockchain is used as a database to store patient billing data, then in a large-scale healthcare scenario, millions of records will be replicated on all Blockchain miners. Moreover, billing data could include detailed patient insurance information, such as their ID number, insurance contact information, coverage details, and other aspects that the provider needs to bill for services. Capturing all these information for every patient generates excessive amounts of data in the Blockchain.

For example, suppose it is necessary to store some insurance and billing information (encrypted) in the Blockchain. In reality, most patients are covered by one of a relatively small subset of insurers (in comparison to the total number of patients, e.g., each insurance policy may cover 10,000s or

100,000s of patients). A substantial amount of nonvarying information, such as details on what procedures are covered by an insurance policy, is common across patients that can therefore be reused and shared. To bill for a service, however, this common intrinsic information must be combined with extrinsic, varying information (such as the patient's ID number and billing address) that is specific to each patient. A good design should maximize sharing of such common data to reduce on-chain storage and, meanwhile, is capable of providing access to complete data objects on-demand.

6.2.2 Solution: Apply the Flyweight Pattern to Minimize Data Storage in the Blockchain

Combining the Flyweight pattern [60] with a factory object can help minimize data storage in the Blockchain. In particular, the factory can establish a registry model that stores shared data between a set of entities in a common contract, i.e., the registry, while externalizing varying data to be stored in entity-specific contracts. The registry can also maintain references (i.e., addresses) to entity-specific contracts and return a combined extrinsic and intrinsic (common) dataset upon request. Fig. 7 shows the flyweight registry model applied to DASH.

In the example above, DASH implements a registry model to manage patient information. Specifically, *PatientFlyweightFactory* is a patient registry that creates a *PatientAccountFlyweight* contract for each patient and links the flyweight reference address to the patient via a unique identifier to avoid account duplication. It stores some data common to different groups of patients (such as insurance coverage as described previously), preventing an exorbitant amount of memory usage from saving repeated data in all patient accounts. *PatientAccountFlyweight* can have concrete implementations, e.g., *ClinicPatientFlyweight* and *HospitalPatientFlyweight* based on the patient's primary care type, to store varying, patient-specific contact, and billing information based. *PatientFlyweightFactory* only create a new account if the specified patient identifier does not yet exist in the registry; otherwise, the registry retrieves the address associated with the existing *PatientAccountFlyweight* contract. To retrieve the complete insurance and billing information of a particular patient, *Client* (DASH server) only needs to invoke a function call from *PatientFlyweightFactory* with the patient identifier to obtain the *PatientAccountFlyweight* address. The function *getData()* of *PatientAccount-Flyweight* is then responsible for returning the combined intrinsic and extrinsic data object back to the *Client*.

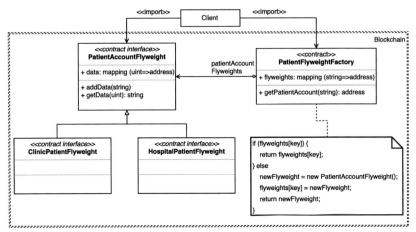

Fig. 7 Applying the Flyweight registry model to create unique patient accounts in DASH design. *Reprinted with permission from P. Zhang, J. White, D.C. Schmidt, G. Lenz, Applying software patterns to address interoperability in Blockchain-based healthcare apps, The 24th Pattern Languages of Programming Conference, October 22–25, 2017, Vancouver, Canada.*

6.2.3 Consequences of Applying the Flyweight Pattern

The flyweight registry model provides better management to the large object pool (e.g., patient accounts in the example above). It minimizes redundancy in similar objects by maximizing data and data operation sharing. Particularly in the above example, if some common insurance policy details were stored in each patient's contract directly, any change to a policy detail would be costly since it would require rewriting a huge number of impacted contracts. Data sharing with flyweight registry can help minimize the cost of changes to the common intrinsic state in Blockchain-based apps. One downside of applying the Flyweight pattern, however, is the added layer of complexity. For example, creating the flyweight contract is another transaction to verify and include in the Blockchain before it can be executed. Although, this extra step can be outweighed by the efficiency in entity/data management that the registry model provides.

6.3 Privacy: Balancing Data Accessibility With Privacy Concerns

6.3.1 Design Problems Faced by Blockchain-Based Apps

If a Blockchain-based healthcare app must expose sensitive data or common meta data (such as patient identifying information or insurance information in the example in Section 5.2) on the Blockchain, it must be designed to maximize health data privacy on-chain while facilitating health information exchange. In particular, a fundamental aspect of a Blockchain is that data and all data change history stored in the Blockchain are public, immutable, and verifiable. For financial transactions focused on proving that the transfer of an asset indeed occurred, these properties are critical. When the goal is to store data in the Blockchain, however, it is important to understand how these properties will impact the use case.

For example, storing patient health records in the Blockchain can be problematic since it requires that data to be public and immutable. Although data can be encrypted before being stored, should all patient data be publicly distributed to all Blockchain nodes? If a consortium Blockchain were deployed among large US hospital organizations with more than 10 hospitals, we would be storing these patient data roughly 40 times [62]. Even if encryption is used, the encryption technique may be broken in the future or bugs in the implementation of the encryption algorithms or protocols used may make the data decryptable in the future. Immutability, however, prevents owners of the data from removing data or its change history from the Blockchain when a security flaw is found.

Many other scenarios, ranging from discovery of medical mistakes in the data to changing standards may necessitate the need to change the data over time. In scenarios where the data may be changed, the public and immutable nature of the Blockchain creates a fundamental tension that must be resolved. On the one hand, healthcare providers would want incorruptible data so it can be trusted and never compromised. At the same time, however, providers may need the data to be changeable so that they can account for possible errors. A practical Blockchain-based health app should protect patient privacy and also ensure data integrity.

6.3.2 Solution: Apply the Proxy Pattern to Enable Secure and Private Data Services

Combining the Proxy pattern [60] with a secure data retrieving service, such as an Oracle [63], can enable secure and private data exchange services. The Oracle network is a third-party service that allows a smart contract to query or retrieve data sources outside the Blockchain address space and ensures that retrieved data are genuine and uncompromised. To reduce computation overhead on-chain, a proxy can be created as a lightweight representation or placeholder for the real data until its retrieval is required. Fig. 8 shows the application of a proxy in DASH.

DASH uses a *Proxy* contract to expose some simple metadata about a patient and later refer to the actual heavyweight implementation *RealPatientData* on-demand to retrieve the complete data object via an

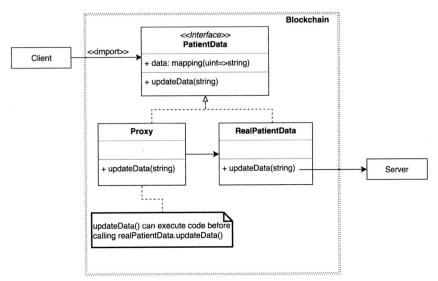

Fig. 8 Structure of a Proxy application in DASH. *Reprinted with permission from P. Zhang, J. White, D.C. Schmidt, G. Lenz, Applying software patterns to address interoperability in Blockchain-based healthcare apps, The 24th Pattern Languages of Programming Conference, October 22–25, 2017, Vancouver, Canada.*

Oracle. Each query request and modification operation are logged in an audit trail that is transparent to the entire Blockchain network for verification against data corruption or unauthorized data access. In the case of a proxified contract (heavyweight implementation) being updated with a new storage option (e.g., replacing an Oracle with some other data service), the interface to the proxy contract can remain unchanged, encapsulating the low-level implementation variations in the proxified contract.

6.3.3 Consequences of Applying the Proxy Pattern

A proxy object can perform lightweight housekeeping or auditing tasks by storing some commonly used metadata in its internal states without having to perform expensive operations (such as retrieving health data via an Oracle service). It typically follows the same interface as the real object and can execute the original heavyweight function implementations as needed. It can also hide information about the real object as needed to protect patient data privacy. However, Proxy may cause disparate behavior when the real object is accessed directly by some *Client* while the proxy surrogate is accessed by others. Nonetheless, proper usage of a proxy with an Oracle service can provide a private channel for protected information exchange.

6.4 Scalability: Tracking Relevant Events Scalably Across Large Traffic

6.4.1 Design Problem Faced by Blockchain-Based Apps

A practical Blockchain-based health system may need to manage and track relevant health events across large patient populations. Therefore, it should be designed to filter out useful health-related information from all communication traffic (i.e., transaction records) occurring on the Blockchain. For example, the Ethereum public Blockchain maintains a transparent record of all contract creation and operation execution history along with regular cryptocurrency transactions. The availability of information makes Blockchain a potentially autonomous approach to improve care coordination across different participants and teams (e.g., physicians, pharmacists, and insurance agents) who would normally communicate through various channels that are manual and time consuming, such as through telephoning or faxing [64].

However, records on the Blockchain are continually growing. Without a meticulously crafted design, capturing any specific health-related topic from all occurred events on-demand would imply exhaustive transaction receipt lookups and topic filtering, which requires nontrivial computations and may result in delayed responses. A good design should support relevant health information relays to facilitate coordinated care as the events occur. For instance, care should be seamless from the point when a patient self-reports illness (e.g., through a familiar web or mobile interface) to the point when the patient receives the necessary prescriptions created by their primary care provider to treat the reported symptoms. Moreover, clinical reports and follow-up procedure should be relayed to and from the associated care centers (e.g., care provider office and pharmacy) in a timely manner.

6.4.2 Solution: Apply the Publisher-Subscriber Pattern to Facilitate Scalable Information Filtering

Incorporating a notification service using the Publisher-Subscriber pattern [61] can facilitate scalable information filtering. In this design, health activities are only broadcast to providers who subscribe to events relating to their patients. It alleviates the tedious filtering process of determining which care provider should be notified about what patient activities as large volumes of transactions take place. This design also fosters an interoperable environment, which allows providers across various organizations or regions to participate. To avoid computation overhead on the Blockchain, the actual processing of patient activities data can be performed off-chain by the inter-facing DApp server. Specifically, when the publisher sends an update, its subscribers only need to do a simple update to an internal state variable that records the publisher's contract address, which the DApp server actively monitors for changes. When a change occurs, the responsibility for the computation–heavy content filtering task (e.g., retrieving the change activity from the publisher using the address) is delegated to the DApp server from the Blockchain. The DApp server is context aware at this point, because each subscriber has an associated contract address accessible by the server. The server can then filter the content based on subscribed topics and update the contract states of appropriate subscribers as needed. Fig. 9 shows the Publisher-Subscriber pattern applied in our DASH design for the notification service.

In DASH, events associated with patient-reported sickness symptoms (a concrete *Topic*) are subscribed to by the patient's primary care provider and pharmacist (i.e., *Clinician* and *Pharmacist* types of *Subscribers*). When this event occurs, the *PatientPublisher* contract notifies the *Subscribers* by updating the *state* variable value in the concrete *Subscriber* contracts. DASH server, which is actively listening for changes in the *Subscriber* states, finds the updated status, and in turn queries the latest patient health changes to pass onto the proper *Subscriber* objects.

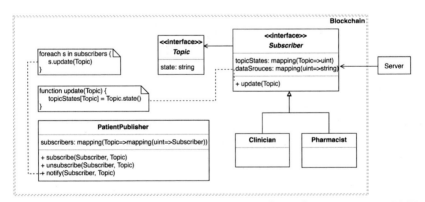

Fig. 9 Using the Publisher-Subscriber pattern to provide notification service in DASH. *Reprinted with permission from P. Zhang, J. White, D.C. Schmidt, G. Lenz, Applying software patterns to address interoperability in Blockchain-based healthcare apps, The 24th Pattern Languages of Programming Conference, October 22–25, 2017, Vancouver, Canada.*

6.4.3 Consequences of Applying the Publisher-Subscriber Pattern

Applying a notification service in a healthcare DApp design is useful when a state change in one contract must be reflected in others without keeping the contracts tightly coupled. Adding or removing subscribers and topics is trivial as it only requires minimal changes in the state variables and not the interface or implementation. It also makes topic/content filtering more manageable when subscription relations are clearly defined in each participant's contract state. In addition, this design enables communication across participants from various organizations, as required of an interoperable system.

However, unlike a traditionally centralized notification service, there are Blockchain-specific limitations, which include:

(1) Potential delays in updates received by subscribers due to the extra step of validation required by the Blockchain infrastructure.

(2) High storage requirements associated with defining very fine-grained topic subscriptions on-chain.

Concretely, in a Blockchain the order of which transactions to be executed and verified is determined by the miners based on some predefined rules. In Blockchains with financial mining incentives, for instance, the priority of a transaction may be based on transaction fees paid by its sender and how long it has been in the transaction pool. Transactions with the highest priority will be executed first and added to the Blockchain sooner, which could cause delays in the notification service for transactions with lower priority. If subscribers want to receive fine-grained message topics, the amount of computation for filtering out the messages sent out by publishers may also cause overhead. This overhead may result in either failure to publish messages due to network limitation or restriction enforced on the computation. One mitigation strategy may be to have broader topics with fewer filter requirements on-chain and to handle more detailed message filtering off the Blockchain.

7. SUMMARY OF KEY DESIGN LESSONS LEARNED

Blockchain and programmable smart contracts provide a platform for creating decentralized systems and applications that may serve a wide range of use cases in the healthcare industry, such as facilitating data sharing, managing patient digital identity, enabling PHRs, tracking prescriptions, expediting claims adjudication, and many more. Properly leveraging Blockchain technologies, given the complexity of the healthcare domain, however, requires that key domain-specific concerns be addressed in the ecosystem design. These concerns include but are not limited to application system evolvability, storage requirements minimization, patient data privacy protection, and application scalability across large user populations. This chapter described these concerns and recommended approaches to mitigate these challenges through an example using our Blockchain-based DApp for Smart Health (DASH). Specifically, we detailed the applications of four familiar software patterns, namely, Abstract Factory, Flyweight, Proxy, and Publisher-Subscriber.

Based on our experience developing the DASH case study presented in this chapter, we learned the following lessons:

- The public, immutable, and verifiable properties of the Blockchain enable a more interoperable environment that is not easily achieved using traditional approaches that mostly rely on a centralized server or data storage.
- Each time a smart contract is modified, a new contract object is created on the Blockchain. Important design decisions must therefore be made in advance to avoid the cost and storage overhead introduced by changes in contract interface.
- To best leverage these properties of Blockchain in the healthcare context, concerns regarding system evolvability, storage costs, sensitive information privacy, and application scalability must be taken into account.
- Combining time-proven design practices with the unique properties of the Blockchain helps guide the design for health systems that are more modular, easier to integrate and maintain, and less susceptible to change.

8. RESEARCH DIRECTIONS IN HEALTHCARE-FOCUSED BLOCKCHAIN APPLICATIONS

There are many research directions in applying Blockchain technology to the healthcare industry due to the complexity of this domain and the need for more robust and effective information technology systems. An interoperable architecture would undoubtedly play a significant role throughout many healthcare use cases that face similar data sharing and communication challenges. From the more technical aspect, much research is needed to pinpoint the most practical design process in creating an interoperable ecosystem using the Blockchain technology while balancing critical security and confidentiality concerns in healthcare. Whether to create a decentralized application leveraging an existing Blockchain, such as Ethereum [3], IBM's Hyperledger Fabric [65], or J.P·Morgan's Quorum [66], or to design a healthcare domain- or use case-specific Blockchain remains an open question.

Additional research on secure and efficient software practice for applying the Blockchain technology in healthcare is also needed to educate software engineers and domain experts on the potential and also limitations of this new technology. Likewise, validation and testing approaches to gage the efficacy of Blockchain-based healthcare architectures compared to existing systems are also important (e.g., via performance metrics related to time and cost of computations or assessment metrics related to its feasibility). In some cases, a new Blockchain network may be more suitable than the existing Blockchains; therefore, another direction may be investigating extensions of an existing Blockchain or creating a healthcare Blockchain that exclusively provides health–related services.

9. CONCLUSION

In this chapter, we described two pressing issues in healthcare, focusing on the need to (1) create an interoperable system to facilitate clinical communications and data exchange and (2) enable patient-centric care to provide patients with access and control of their complete medical history. We then identified seven concrete healthcare scenarios that share similar technical pain points, which can be alleviated with blockchain technology. However, the complexities associated with healthcare involvement and regulations create additional challenges inevitably facing blockchain-based systems, such as system evolvability, information privacy, and communication scalability. Targeting a subset of these healthcare-specific challenges, we presented four design recommendations demonstrated with a case study prototype that we have previously developed. From our experience, we have seen the great potential of blockchain technology in creating secure and effective healthcare ecosystems with its inherent unique properties. In addition, we have also observed the importance of integrating domain-specific concerns and need into blockchain-based designs. Overall, blockchain has a wide range of possibilities in healthcare, which invites many research opportunities in this space.

KEY TERMINOLOGY AND DEFINITIONS[c]

Abstract Factory Pattern Abstract Factory encapsulates a group of individual factories that have a common theme without specifying their concrete classes. It separates the details of implementation of a set of objects from their general usage and relies on object composition, as object creation is implemented in methods exposed in the factory interface.

Cryptocurrency A cryptocurrency is a digital or virtual currency that uses cryptography for security, making it difficult to counterfeit. It is not issued by any central authority and is immune to government interference or manipulation.

Decentralized App (DApp) A DApp is an autonomously operated open-source application that cryptographically stores its data and records of operation in a public, decentralized Blockchain (via a smart contract for instance) to avoid central points of failure. It uses a native or an existing form of cryptographic tokens for monetizing the DApp. The tokens must be necessary for the use of the app.

Digital Asset A digital asset is anything existing in a binary format that comes with (some) rights to use. It could be a native asset lacking physical substance that can be owned or controlled to produce value, such as digital music, images, movies, electronic funds, and software. It could also be a digital representation of some traditional paper-based asset, such as certificates and titles of property, gold, autos, stock, and currency.

Electronic Health Records (EHRs) EHRs are a digital version of a patient's paper medical records and chart that make information available instantly and securely to authorized healthcare practitioners. They contain the medical and treatment histories of patients and can also store information beyond standard clinical data collected in a provider's office, such as diagnoses, medications, treatment plans, allergies, and lab results.

Factory A factory is a function or method that creates an object of a varying prototype or class from some method call. It creates abstraction or encapsulation so that program code is not tied to specific classes or objects, allowing the class hierarchy or prototypes to be changed or refactored without modifying code that uses them.

Flyweight Pattern A flyweight object minimizes memory usage by sharing as much data as possible with other similar objects. It is particularly when a simple repeated representation would use an unacceptable amount of memory. Common parts of the object state can be shared internally, while varying parts of the data are stored externally in entity-specific objects. When all data regarding a specific object are requested, both external and internal data can be retrieved.

Interoperability Interoperability allows two or more systems to exchange information and use the exchanged information. The three levels of health information technology interoperability ordered from the lowest to the highest fidelity are: (1) *Foundational* interoperability that enables data exchanges between healthcare systems without requiring the ability for the receiving party to interpret the data, (2) *Structural* interoperability that defines the formats of exchanged clinical data and ensures that received data are preserved and can be interpretable at the data field using the predefined formats, and (3) *Semantic* interoperability that allows for interpretation of data exchanged by not only syntax (structure) but also semantics (meaning) of the data.

---

[c]Key technology/technical terms used in this chapter will be explained wherever they appear or at the "Key Terminology and Definitions" section. Apart from regular References, additional References are included in the "References for Advance/Further reading" for the benefit of advanced readers.

Master Patient Index (MPI) Master patient index is a single registration system of all patients across various departments within a hospital. Similarly, an enterprise master patient index (EMPI) is a database that consolidates patient identities from multiple healthcare organizations. The goal of MPI or EMPI is to provide uniquely identifying patient information so that it may be queried and used to match existing records.

ProxyPattern A proxy is a wrapper object that is used by the client to access the real serving object behind the scenes. It implements the same interface as the real object and can execute the original heavyweight function implementations as needed. A proxy can provide extra functionality that is typically lightweight housekeeping or auditing tasks, such as checking preconditions or caching when operations on the real object are resource intensive.

Publisher–Subscriber Pattern This is a messaging pattern where senders of messages, publishers, do not directly send to specific receivers, called subscribers. Instead, publishers categorize published messages into topics without the knowledge of which subscribers. Subscribers express interest in one or more topics and only receive messages that are of interest, without knowledge of which publishers.

Smart Contract Smart contracts are enhancements built atop some Blockchain technologies (such as Ethereum). They are code that directly controls the exchanges or redistributions of digital assets between two or more parties according to certain rules or agreements established between the involved parties. They enable development of DApps to interact with Blockchain and support on-chain storage.

Software Pattern A software pattern is a general repeatable solution to a commonly occurring problem in software design. It is not a finished design that can be transformed directly into code. Instead, it provides a description or template for how to solve a problem that can be used in many different situations. Software patterns allow developers to communicate using well-known, well-understood names for software interactions. Common design patterns can be improved over time, making them more robust than ad hoc designs.

REFERENCES

[1] S. Nakamoto, Bitcoin: A Peer-to-Peer Electronic Cash System, 2008. https://bitcoin.org/bitcoin.pdf.

[2] D. Johnston, S.O. Yilmaz, J. Kandah, N. Bentenitis, F. Hashemi, R. Gross, S. Wilkinson, S. Mason, The General Theory of Decentralized Applications, DApps, GitHub, 2014. 9.

[3] V. Buterin, Ethereum White Paper, GitHub, 2013. GitHub repository.

[4] P. Zhang, J. White, D.C. Schmidt, G. Lenz, Design of blockchain-based apps using familiar software patterns with a healthcare focus, in: the 24th Pattern Languages of Programming Conference, October 22–25, Vancouver, Canada, 2017.

[5] K.B. DeSalvo, RE: Connecting Health and Care for the Nation: A Shared Nationwide Interoperability Roadmap, 2015. https://www.healthit.gov/policy-researchers-implementers/interoperabilityPublished on healthit.gov.

[6] S. Olson, A. Downey, Sharing clinical research data: workshop summary. 2013, in: A.S. Downey, S. Olson (Eds.), Book Sharing Clinical Research Data: Workshop Summary. 2013, National Academies Press, Washington, DC, 2016.

[7] The Biggest Healthcare Breaches of 2017, Retrieved March 01, 2018, from, http://www.healthcareitnews.com/slideshow/biggest-healthcare-breaches-2017-so-far?page=18, 2017.

[8] J. Adler-Milstein, D.W. Bates, Paperless healthcare: progress and challenges of an IT-enabled healthcare system, Bus. Horiz. 53 (2) (2010) 119–130.

[9] P. Zhang, M. Walker, J. White, D.C. Schmidt, G. Lenz, in: Metrics for assessing Blockchain-based healthcare decentralized apps, Proceedings of 2017 IEEE 19th International Conference on e-Health Networking, Applications and Services (Healthcom), October 12–15, Dalian, China, 2017.

[10] R.H. Dolin, L. Alschuler, C. Beebe, P.V. Biron, S.L. Boyer, D. Essin, E. Kimber, T. Lincoln, J.E. Mattison, The HL7 clinical document architecture, J. Am. Med. Inform. Assoc. 8 (6) (2001) 552–569.

[11] D. Bender, K. Sartipi, HL7 FHIR: an agile and RESTful approach to healthcare information exchange, in: Computer-Based Medical Systems (CBMS), 2013 IEEE 26th International Symposium on. IEEE, 2013, pp. 326–331.

[12] P.P. Reid, W.D. Compton, J.H. Grossman, G. Fanjiang, Building a Better Delivery System: A New Engineering/Health Care Partnership, National Academies Press, Washington, DC, 2005.

[13] B. Middleton, M. Bloomrosen, M.A. Dente, B. Hashmat, R. Koppel, J.M. Overhage, T.H. Payne, S.T. Rosenbloom, C. Weaver, J. Zhang, Enhancing patient safety and quality of care by improving the usability of electronic health record systems: recommendations from AMIA, J. Am. Med. Inform. Assoc. 20 (e1) (2013) e2–e8.

[14] C. Castaneda, K. Nalley, C. Mannion, P. Bhattacharyya, P. Blake, A. Pecora, A. Goy, K.S. Suh, Clinical decision support systems for improving diagnostic accuracy and achieving precision medicine, J. Clin. Bioinforma. 5 (1) (2015) 4.

[15] H. Singh, T.D. Giardina, A.N. Meyer, S.N. Forjuoh, M.D. Reis, E.J. Thomas, Types and origins of diagnostic errors in primary care settings, JAMA Intern. Med. 173 (6) (2013) 418–425.

[16] G.D. Schiff, O. Hasan, S. Kim, R. Abrams, K. Cosby, B.L. Lambert, A.S. Elstein, S. Hasler, M.L. Kabongo, N. Krosnjar, Diagnostic error in medicine: analysis of 583 physician-reported errors, Arch. Intern. Med. 169 (20) (2009) 1881–1887.

[17] R. Kaushal, K.G. Shojania, D.W. Bates, Effects of computerized physician order entry and clinical decision support systems on medication safety: a systematic review, Arch. Intern. Med. 163 (12) (2003) 1409–1416.

[18] E. Warren, Strengthening research through data sharing, N. Engl. J. Med. 375 (5) (2016) 401–403.

[19] D.B. Taichman, J. Backus, C. Baethge, H. Bauchner, P.W. De Leeuw, J.M. Drazen, J. Fletcher, F.A. Frizelle, T. Groves, A. Haileamlak, Sharing clinical trial data—a proposal from the International Committee of Medical Journal Editors, N. Engl. J. Med. 374 (2016) 384–386.

[20] N. Geifman, J. Bollyky, S. Bhattacharya, A.J. Butte, Opening clinical trial data: are the voluntary data-sharing portals enough? BMC Med. 13 (1) (2015) 280.

[21] G.E. Gross, The role of the tumor board in a community hospital, CA Cancer J. Clin. 37 (2) (1987) 88–92.

[22] Nourie, C. E. (Ed.). (2015). Your Medical Records. Retrieved March 01, 2018, from http://m.kidshealth.org/en/teens/medical-records.html.

[23] R. Schoenberg, Bridged patient/provider centric method and system, U.S. Patent No. 8,457,981, 2013.

[24] J. Lumpkin, S.P. Cohn, J.S. Blair, Uniform Data Standards for Patient Medical Record Information, National Committee on Vital and Health Statistics, 2003. 53.

[25] S. Heath, How is Interoperability Supporting Patient-Centered Care?, Retrieved March 01, 2018, from, https://ehrintelligence.com/news/how-is-interoperability-supporting-patient-centered-care, 2016.

[26] N. Black, M. Varaganum, A. Hutchings, Relationship between patient reported experience (PREMs) and patient reported outcomes (PROMs) in elective surgery, BMJ Qual. Saf. 23 (2014) 534–542. bmjqs-2013-002707.

[27] J.S. Ash, M. Berg, E. Coiera, Some unintended consequences of information technology in health care: the nature of patient care information system-related errors, J. Am. Med. Inform. Assoc. 11 (2) (2004) 104–112.

[28] G. Hripcsak, M. Bloomrosen, P. FlatelyBrennan, C.G. Chute, J. Cimino, D.E. Detmer, M. Edmunds, P.J. Embi, M.M. Goldstein, W.E. Hammond, Health data use, stewardship, and governance: ongoing gaps and challenges: a report from AMIA's 2012 Health Policy Meeting, J. Am. Med. Inform. Assoc. 21 (2) (2014) 204–211.

[29] R. LaRose, S. Strover, J.L. Gregg, J. Straubhaar, The impact of rural broadband development: lessons from a natural field experiment, Gov. Inf. Q. 28 (1) (2011) 91–100.

[30] L. Manchikanti, S. Helm, B. Fellows, J.W. Janata, V. Pampati, J.S. Grider, M.V. Boswell, Opioid epidemic in the United States, Pain Physician 15 (2012). ES9–38.

[31] I. Bernstein, Drug Supply Chain Security Act, in: Presented at: FDLI DQSA Conference, 2017.

[32] G.C. Christie, C.G.C. Baker, G.R. Cooper, C.P.J. Kennedy, B. Madras, F.A.G.P. Bondi, The President's Commission on Combating Drug Addiction and the Opioid Crisis, Homeland Security Digital Library, 2018. https://www.hsdl.org/?abstract&did=805384.

[33] CDC Rx Awareness Campaign Overview, Retrieved March 1, 2018, from, https://www.cdc.gov/rxawareness/pdf/RxAwareness-Campaign-Overview-a.pdf, 2016.

[34] C. Arlotta, Opioid Overdose and Abuse Awareness Campaigns Continue to Grow, Retrieved March 01, 2018, from, https://www.forbes.com/forbes/welcome/?toURL=https%3A%2F%2Fwww.forbes.com%2Fsites%2Fcjarlotta%2F2015%2F07%2F31%2Fopioid-overdose-abuse-awareness-campaigns-continue-to-grow%2F&refURL=https%3A%2F%2Fwww.google.com%2F&referrer=https%3A%2F%2Fwww.google, 2015.

[35] J. James, Dealing with drug-seeking behaviour, Aust. Prescr. 39 (3) (2016) 96.

[36] L.S. Nelson, D.N. Juurlink, J. Perrone, Addressing the opioid epidemic, JAMA 314 (14) (2015) 1453–1454.

[37] 10 Pros and Cons of Telemedicine | eVisit® Telehealth Solutions, Retrieved March 01, 2018, from, https://evisit.com/10-pros-and-cons-of-telemedicine/, 2018.

[38] S. Sood, V. Mbarika, S. Jugoo, R. Dookhy, C.R. Doarn, N. Prakash, R.C. Merrell, What is telemedicine? A collection of 104 peer-reviewed perspectives and theoretical underpinnings, Telemed. J. eHealth 13 (5) (2007) 573–590.

[39] What Is Telemedicine? Definition From WhatIs.com, Retrieved March 01, 2018, from, http://searchhealthit.techtarget.com/definition/telemedicine, 2016.

[40] IOS—Health, Retrieved March 01, 2018, from, https://www.apple.com/ios/health/, 2018.

[41] P. Zhang, J. White, D.C. Schmidt, G. Lenz, S.T. Rosenbloom, FHIRChain: Applying Blockchain to Securely and Scalably Share Clinical Data, 2018 (preprint).

[42] Multidisciplinary Cancer Care: The Benefits of a Tumor Board, Retrieved March 01, 2018, from, https://www.maacenter.org/blog/multidisciplinary-cancer-care-the-benefits-of-a-tumor-board, 2017.

[43] M. Wanner, Why is Cancer So Difficult to Cure?, Retrieved March 01, 2018, from, https://www.jax.org/news-and-insights/2015/december/why-no-cure-for-cancer, 2015.

[44] Cancer Patients Face the Ultimate Choice, With No Room for Error, Retrieved March 01, 2018, from, https://www.statnews.com/2016/10/06/immunotherapy-cancer-clinical-trials/, 2017.

[45] D.M. Parkin, The evolution of the population-based cancer registry, Nat. Rev. Cancer 6 (8) (2006) 603.

[46] B.H. Just, D. Marc, M. Munns, R. Sandefer, Why patient matching is a challenge: research on master patient index (MPI) data discrepancies in key identifying fields, Perspect. Health Inf. Manag. 13 (Spring) (2016). https://www.ncbi.nlm.nih.gov/pmc/articles/PMC4832129/.

[47] C. Feied, F. Iskandar, Master patient index, U.S. Patent Application No. 11/683,799, 2007.

[48] Cross Organizational Patient Identity Management: Challenges and Opportunities, Retrieved March 1, 2018, from, http://sequoiaproject.org/wp-content/uploads/2017/02/2017-02-22-HIMSS-2017-Patient-Matching-Challenges-and-Opportunities-v001.pdf, 2016.

[49] J.L. Fernández-Alemán, I.C. Señor, P.Á.O. Lozoya, A. Toval, Security and privacy in electronic health records: a systematic literature review, J. Biomed. Inform. 46 (3) (2013) 541–562.

[50] Patient Matching Errors Risk Safety Issues, Raise Health Care Costs, Retrieved March 1, 2018, from, http://www.pewtrusts.org/en/multimedia/data-visualizations/2017/patient-matching-errors-risk-safety-issues-raise-health-care-costs, 2017.

[51] Patient Matching Peril: Why Unique Patient Identifiers are a Unique Problem for Hospitals, Retrieved March 01, 2018, from, https://www.beckershospitalreview.com/healthcare-information-technology/patient-matching-peril-why-unique-patient-identifiers-are-a-unique-problem-for-hospitals.html, 2016.

[52] R. Krawiec, D. Housman, M. White, M. Filipova, F. Quarre, D. Barr, A. Nesbitt, K. Fedosova, J. Killmeyer, A. Israel, Blockchain: opportunities for health care, in: Proc. NIST Workshop Blockchain Healthcare, 2016, pp. 1–16.

[53] P.C. Tang, J.S. Ash, D.W. Bates, J.M. Overhage, D.Z. Sands, Personal health records: definitions, benefits, and strategies for overcoming barriers to adoption, J. Am. Med. Inform. Assoc. 13 (2) (2006) 121–126.

[54] Microsoft HealthVault, Retrieved March 01, 2018, from, https://www.healthvault.com/en-us/, 2017.

[55] What Is Health Insurance, Retrieved March 1, 2018, from, https://www.medicalnewstoday.com/info/health-insurance, 2016.

[56] B.E. Peterson, J.W. Kwant Jr, V.C. Cecil, W.A. Provost, Electronic creation, submission, adjudication, and payment of health insurance claims, U.S. Patent No. 6,343,271, 2002.

[57] Claims Processing: What Is Claims Adjudication?, Retrieved March 01, 2018, from, http://www.apexedi.com/what-is-claims-adjudication/, 2017.

[58] Blockchain's Potential Use Cases for Healthcare: Hype or Reality?, Retrieved March 01, 2018, from, http://www.healthcareitnews.com/news/blockchains-potential-use-cases-healthcare-hype-or-reality, 2017.

[59] 5 Key Challenges in Medical Billing Industry, Retrieved March 01, 2018, from, http://www.invensis.net/blog/infographics/5-key-challenges-in-medical-billing-industry/, 2017.

[60] E. Gamma, Design Patterns: Elements of Reusable Object-Oriented Software, Pearson Education India, 19951995.

[61] F. Buschmann, K. Henney, D. Schimdt, Pattern-Oriented Software Architecture: On Patterns and Pattern Language, John Wiley & Sons, 2007, 2007.

[62] Top 30 Largest Hospital Systems in America, Retrieved March 01, 2018, from, http://www.compassphs.com/blog/healthcare-trends/healthcare-fast-facts-top-30-largest-hospital-systems-in-america/, 2017.

[63] M. Swan, Blockchain thinking: The brain as a dac (decentralized autonomous organization), in: Texas Bitcoin Conference, Chicago, 2015, pp. 27–29.
[64] EHRs and Healthcare Interoperability: The Challenges, Complexities, Opportunities and Reality, Retrieved March 01, 2018, from, http://www.healthcareitnews.com/blog/ehrs-healthcare-interoperability-challenges-complexities-opportunities-reality, 2015.
[65] Hyperledger Fabric, Retrieved March 01, 2018, from, http://www.hyperledger.org/projects/fabric, 2018.
[66] Quorum, Retrieved March 01, 2018, from, https://www.jpmorgan.com/global/Quorum, 2017.

ABOUT THE AUTHORS

Peng Zhang is a Computer Science PhD student at Vanderbilt University, Nashville, TN, USA. Her research interests include model-driven design for engineering and healthcare IT systems, intelligent model constructions using machine and deep learning, decentralized algorithms and protocols for facilitating and securing clinical communications, and application and enhancement of Blockchain technologies for moving toward patient-centered care.

Douglas C. Schmidt is Cornelius Vanderbilt Professor of Computer Science, the Associate Chair of the Electrical Engineering and Computer Science Department, and a Senior Researcher at the Institute for Software Integrated Systems, all at Vanderbilt University. He is also a Visiting Scientist at the Software Engineering Institute (SEI) at Carnegie Mellon University. As an internationally renowned and widely cited (h-index of 80) researcher, his work focuses on patterns, optimization techniques, and empirical analyses of object-oriented and component-based frameworks and model-driven engineering tools that facilitate the development of distributed real-time and embedded (DRE) middleware frameworks and mobile cloud computing applications on parallel platforms running over wireless/wired networks and embedded system interconnects. He has published over 10 books and 600+ papers (including 100+ journal papers) in top IEEE, ACM, IFIP, and USENIX technical journals, conferences, and books that cover a range of topics. He has mentored and graduated over 40 PhD and Masters students working on these research topics and has presented 550+ keynote addresses, invited talks, and tutorials on mobile cloud computing with Android, reusable patterns, concurrent object-oriented network programming, and distributed system middleware at scores of technical conferences.

Jules White is an Assistant Professor in the Department of Electrical Engineering and Computer Science at Vanderbilt University. He was previously a Faculty member in Electrical and Computer Engineering and won the Outstanding New Assistant Professor Award both at Virginia Tech. His research has published over 120 papers and won 5 Best Paper and Best Student Paper Awards. His research focuses on securing, optimizing, and leveraging data from mobile cyber-physical systems. His mobile cyber-physical systems research spans focus on: (1) mobile security and data collection, (2) high-precision mobile augmented reality, (3) mobile device and supporting cloud infrastructure power and configuration optimization, and (4) applications of mobile cyber-physical systems in multidisciplinary domains, including energy-optimized cloud computing, smart grid systems, healthcare/manufacturing security, next-generation construction technologies, and citizen science. His research has been licensed and transitioned to industry, where it won an Innovation Award at CES 2013, attended by over 150,000 people, was a finalist for the Technical Achievement at Award at SXSW Interactive, and was a top three for mobile in the Accelerator Awards at SXSW 2013. His research is conducted through the Mobile Application computinG, optimizatoN, and secUrity Methods (MAGNUM) Group at Vanderbilt, which he directs.

Gunther Lenz is an influential Software Solution Architect/Executive, MBA, published author, invited speaker, and strategic technology visionary with 16+ years of achievements to drive successfully drive software innovation projects. Established technology, process, and organizational transformation agent. Recognized expert in software architecture, big data, and cloud computing. Diverse experience in enterprise, high-growth, and incubation environments within a variety of industries. Published two books and numerous novel articles in the area of software engineering and software architecture. Selected Program Committee Member at prestigious international conferences. Awarded Microsoft Most Valuable Professional for Software Architecture (180 worldwide) 2005–08.

CHAPTER TWO

Blockchain for a Trust Network Among Intelligent Vehicles

Shiho Kim
School of Integrated Technology, Yonsei University, Seoul, South Korea

Contents

Abstract

The intelligent vehicle communication network is prone to cyberthreats, which are difficult to solve using traditional centralized security approaches. Blockchain is an immutable peer-to-peer distributed database containing cryptographically secured information. Blockchain shows successful use cases in financial applications, smart contact, protecting digital copyright of media contents. It extends to all industries including the secure IoT devices, embedded systems, etc. The superior feature of blockchain is its decentralized, immutable, auditable database that secures transactions by protecting privacy. In this chapter, we contemplate the environment of the intelligent vehicle communication network and issues regarding methods of building a blockchain-based trust network among intelligent vehicles. We present the use cases of blockchain in intelligent vehicles in the phase of ongoing research or that under development from automotive industries and academic institutes. We also deliberate the challenging issue of blockchain for intelligent vehicles.

Advances in Computers, Volume 111
ISSN 0065-2458
https://doi.org/10.1016/bs.adcom.2018.03.010
43

1. INTRODUCTION

Vehicle technology has advanced rapidly toward the ultimate goals of autonomous driverless cars for efficient traffic flow, preventing human error-related accidents, and improving fuel economy, and emerging electric vehicle charging services. Wireless internet access combined with other communication elements is becoming basic features of intelligent vehicles. The modern intelligent connected car has become a computer system similar to a smartphone, controlled by a complex computer with a wireless communication network and internal in-vehicle network. The intelligent vehicles connected to almost everything are evolving rapidly as a cyberphysical system from the traditional rich mechanical system.

However, the introduction of a vehicular cloud network based on wireless internet connectivity means a remarkable increase in the reliance on the networked software. This leads to a cybersecurity issue, as well as safety-related function problems in the automotive industry. Based on these aforementioned facts, one of the most important concerns associated with intelligent vehicles is to use a trust network to secure them against malicious cyberattacks.

Conventional centralized security system is vulnerable because if a central system is compromised, owing to fraud, cyberattack, or a simple error, the entire network is affected. Conventional security and privacy methods used in personal computers or cloud networks tend to be ineffective in intelligent vehicles owing to their vulnerability in cyberattacks.

Blockchain technology works for the cryptocurrency, which has recently been used to build trust and reliability in peer-to-peer networks with similar topologies as networks of intelligent vehicles. Blockchain technology works for cryptocurrency, which has recently been used to build trust and reliability in peer-to-peer networks with similar topologies as networks of intelligent vehicles. Blockchain has shown successful use cases in financial applications, smart contact, and delivering media content with digital copyright protection. Applicable fields have now been extended to secure internet of things (IoT) devices, embedded systems, and to industries other than information-related fields [1]. A superior feature of blockchain is its decentralized, immutable, auditable database for secure transaction with privacy protection [2]. In this chapter, we contemplate the trusted environment of intelligent vehicle communication network and how the blockchain technology creates trust network among intelligent vehicles.

The blockchain technology is gaining momentum and creating new business opportunities in both the automotive and ICT (information and communication) industries to untangle the potential cybersecurity threat of intelligent connected vehicles; however, challenges lie ahead [3]. In order to apply blockchain to intelligent connected vehicles, participants of the vehicle cloud need to overcome a number of technical challenges due to the limited capability of in-vehicle power and storage, and many other factors.

Autonomous vehicles are no longer privately owned assets, but they may create many new business models. Blockchain technology has already disrupted the financial, media content service, and healthcare industries, and the automotive industry will follow [4,5]. Some automakers and research institutes have started to explore blockchain to create new ecosystems of new service and business models for autonomous and connected vehicles [6].

In this chapter, we introduce the key aspects of cybersecurity issues in intelligent vehicles and discuss how we build secure trust networks using blockchain technology.

Rest of the chapter is divided into two sections. In Section 3, we investigate use cases of blockchain in intelligent vehicle proposals and researches initiated from both automotive industries and academic institutes. To apply blockchain to intelligent connected vehicles, participants in the vehicle cloud need to overcome a number of technical challenges due to the limited capability of in-vehicle power and storage, and many other factors. In Section 4, we present challenging issues of blockchain for intelligent vehicles. The technology/technical terms used in this chapter are explained wherever they appear or at the "Key Terminology and Definitions" section. Apart from regular References additional references are included in the "References for Advance/Further Reading" for the benefit of advanced readers.

2. INTRODUCTION TO CYBERSECURITY ISSUES OF INTELLIGENT VEHICLES

Modern intelligent vehicles are controlled by complex computer with connectivity of both internal in-vehicle network and wireless communication network. While this computerized structure has offered a prominent technology for intelligent functionality such as advanced driver-assistant system or self-driving autonomous vehicles, but also it makes the vehicles insecure and potentially vulnerable to various kinds of malicious hackings and cyberattacks. This section presents a brief introduction and use cases of intelligent vehicles and connected vehicles, and then we raise the open question of the cybersecurity issue of intelligent vehicles.

2.1 Brief Introduction to Intelligent Transportation System and Connected Vehicles

We generally define the connected car which is equipped with wireless internet connectivity and usually also with VANET (vehicular ad hoc NETwork). Advances in software and hardware technology lead to vehicles under full computerized intelligent control. Fig. 1 shows a typical architecture of an intelligent connected vehicle. Intelligent vehicles include numerous networked control units with sensors ensuring multiple vehicle functionalities. The necessary hardware devices realizing intelligent and connected vehicle function can be classified into built-in or brought-in connection systems.

Connected cars have become a smartphone-like computer system with internet connection, while it has more complex control through in-vehicle network for the control of motion dynamics, prolusion and drive train, and electronic accessories associated with a variety of functions providing comfort and convenience to occupants.

Sometimes, VANET is referred to as the wireless network known as intelligent transportation system (ITS), which is aiming to enable safe, efficient flow of traffic, environmentally conscious intelligent transportation services, as well as additional traveler information services. ITS is designed

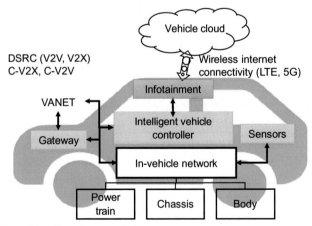

Fig. 1 Outline of intelligent connected car.

to support the car to share or to exchange information among vehicles and infrastructures through the communication protocols such as vehicle-to-vehicle (V2V), vehicle-to-(roadside) infrastructure (V2I), and vehicle-to-network (V2N) or vehicle-to-everything (V2X) communications. The standard protocol used for ITS includes traditional DSRC (dedicated short range communication) or emerging cellular V2X (C-V2X) communication. DSRC, which is based on a family of standards referred to as Wireless Access in Vehicular Environments, is a well-established technology undergone extensive product development and field trials by numerous stakeholders over nearly two decades. In contrast, C-V2X is an emerging technology developed within 5G LTE (long-term evolution) cellular communication by the standardization consortium of 3GPP (the 3rd Generation Partnership Project) [7].

Built-in telematics boxes most commonly used for internet connection via a WIFI and LTE modules integrated in the infotainment (i.e., information plus entertainment) system. The brought-in devices are plugged into the gateway of in-vehicle bus through the on-board diagnostics port, which is a network connector of the scanner and diagnostic tools of vehicles.

Autonomous vehicle is the state-of-the-art of systems where actuation and decision-making maneuvers operate concurrently with coping with incoming environmental information from sensors and communication networks during driving.

Intelligent vehicles will be connected to almost everything, and they will become a part of "IoT." The introduction of a wireless internet connectivity-based vehicle cloud will trigger a remarkable increase in the reliance on the networked software, consequently, leading to issues of cyberthreats for data security, leakage of privacy, as well as safety-related functions in vehicles.

2.2 Use Case/Application of Intelligent Vehicles

This section presents a broad overview of the current potential applications and those under development, as well as use cases of intelligent vehicles. Intelligent connected vehicles communicate with almost all systems and devices in cyberspace, and even those existing at off-line, including smartphones, other vehicles, surrounding roadside infrastructure, various servers, and even offices and homes. Vehicles that become a part of the IoT enable the development of numerous applications offering new services or business opportunities in the public transport, industrial, and entertainment sectors in addition to significant enhancements to safe, efficient, and comfort-driving functions. Fig. 2 shows

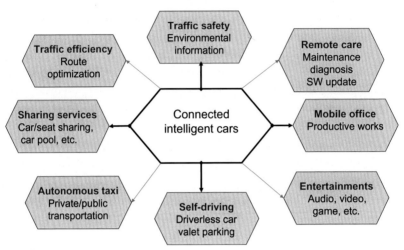

Fig. 2 Service and applications of next-generation intelligent connected cars.

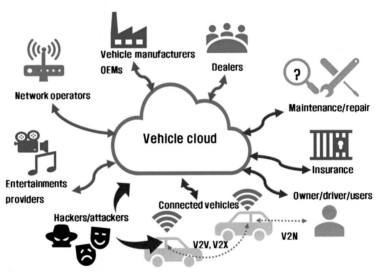

Fig. 3 Ecosystem of the intelligent vehicle industry based on the vehicle cloud.

service and applications of next-generation intelligent connected cars, and
Fig. 3 illustrates an emerging ecosystem of the intelligent vehicle industry
based on the Vehicle Cloud. Key features of current and potential applications
and use cases of the connected vehicle include:

i. *Traffic Safety*: Statistics reveal that more than 90% of traffic accidents
are caused by human error of drivers. Intelligent vehicle will clearly

improve traffic safety and effectively reduce the number of fatalities and injuries by implementing safety-assistant functions that exploit environmental information from in-vehicle sensors and installed communication devices.

ii. *Traffic Efficiency*: Optimization of the route by taking into account data of environmental and traffic condition offers substantial benefits in traffic efficiency by improving traffic flows and fuel efficiency, and reducing travel time.

iii. *Self-Driving/Driverless car*. Advanced intelligent connected cars will be born to support autonomous self-driving without human's involvement in maneuvering; these driverless cars offer people a new realm of experience and service in both private and public transportation. The self-driving cars may be a destructive creation, upsetting established industries and reshaping cities, and providing new options for what do while riding a car. Some examples of simple derived applications are automatic valet parking services, and remote charging services of electric vehicles.

iv. *Sharing Services*: Advancement in vehicle technology allows users to share of facility of transportation facilities more conveniently, and a progress in a new sharing service is ready to come. Established car sharing services still need a driver; however, autonomous cars will reshape the sharing services, to include not only sharing of a car, but also sharing a part of seats or cargo space of vehicles. The implication of a future of shared, autonomous vehicles is that ownership of private car is likely to be disrupted.

v. *Autonomous Taxi*: Traditional mobility operates in two ways: Public transportation at low maintenance cost but for low convenience and Private ownership at a high cost but high convenience. With the advent of on-demand autonomous mobility services, consumers have benefited from lower costs and high convenience. Self-driving cars lead to autonomous taxis, which is a new way of mobility combining the low management cost of public transportation and the convenience of personal mobility demand.

vi. *Entertainments*: Car entertainment systems are a collection of hardware and software in automobiles that provides audio, video or game, and entertainment. Traditional in-car information and entertainment originated with car audio and navigation systems, and this has advanced to include video players, games, and other means of advanced contents for entertainment connected to the internet.

vii. *Remote Care and Maintenance*: Advancements in the technology of wireless connectivity and in-vehicle networked embedded system with sensors enable remote diagnosis and maintenance of vehicles. Over-the-air update is a common way of distributing new software or updating revised version of software in mobile phones. Intelligent vehicles allow the automotive industry to embrace the Over-the-air update technologies, enabling secure remote software updates to be downloaded to the vehicle from a server.

viii. *Mobile Office*: Autonomous vehicle free drivers from having to pay attention to driving. Instead, they are able to rest or conduct productive tasks. Self-driving cars may ultimately morph into a fully connected mobile office so users can work on simple or professional tasks during their commute.

Methods of providing various services, such as the continuous evolution of services, management of protocol, and remote upgrading of vehicle firmware, to vehicle users are a challenge. The advent of wireless internet connectivity enables the realization of a vehicle cloud network that supports intelligent vehicle applications and services based on the cloud computing technique [8,9]. By integrating a bidirectional communication link in the cloud computing technique, convenient customized services and access control can be implemented for vehicle users, service providers, and various parties related to intelligent vehicles.

2.3 Cybersecurity Issue of Intelligent Vehicles

Wireless internet access combined with other communication elements is becoming basic features of intelligent vehicles. Intelligent vehicles are evolving rapidly as a cyber-physical system from the traditional rich mechanical system. Connected intelligent vehicles offer a range of sophisticated services that benefit the vehicle drivers, users, owners, authorities of transport management, automanufacturers, and other service providers. Whenever a device is connected to the internet it is exposed to a risk from various malicious cyberattacks. Advanced cars equipped with digital technology have the most computerized features and are networked to communicate with each other, which opens up new roads for hacking. Like the other ICT industries that are often targets of data breaches, the intelligent vehicles must consider the many potential attack surfaces that hackers can use to break into a network. In addition to the internet connections, other communication elements of automobile are vulnerable to security threats

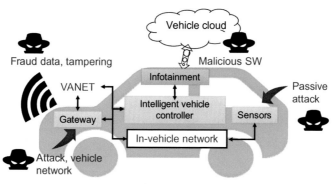

Fig. 4 Potential attack surface of intelligent connected car.

against vehicle safety and the protection of sensitive privacy information, which may lead to fatal consequences. The attack surfaces of intelligent connected car are broad and may extend far beyond the vulnerability of hacking the vehicle itself.

Potential security threats and attack surface of intelligent connected cars can be classified into following four categories as shown in Fig. 4: remote attack through VANET, cyberattack through the internet, passive attack by tampering with sensors data, and attempted unauthorized connection to the in-vehicle network architecture.

i. *Threat of V2X communication*: V2X communications can impose some serious threats against security and safety of vehicles and can lead to disturbed traffic flow. Intelligent vehicles are required to gather information such as the traffic environment, instantaneous traffic situations at invisible distances, and data for cooperative driving. V2X communications are vulnerable to attacks by transmitting fake information or tampering with messages. One of the challenges of blockchain technology for intelligent vehicles is the provision of a secure V2X communication network to trust messages from other vehicles or infrastructures, as well as without privacy leakages.

ii. *Passive attack:* A malicious attack that makes vehicle senses an incorrect information by counterfeiting traffic signals or road signs or by providing fake signals.

iii. *Attempting direct attacks on the in-vehicle network*: An attacker can attempt to make an unauthorized connection to the in-vehicle network, whereby we are able to control the core functions of the vehicle through the gateway of the in-vehicle network.

iv. *Internet connection and/or vehicular cloud:* Internet connection turns intelligent vehicles into a cyber-physical system and each vehicle behaves as a kind of nodes in the IoT [8]. Then, the vehicles are revealed to the potential malicious attackers through various channels by which to access the vehicle's delicate controller or in-vehicle network bus. This potentially exposes intelligent vehicles to a wide range of safety, security, and privacy threats such as malware attacks, location tracking, unauthorized connections, and remote hijacking of the vehicle. This cyberthreat is the biggest risk for autonomous self-driving vehicles. Malicious entities can compromise a vehicle, which not only endangers the security of the vehicle, but also the safety of the passengers.

Based on the aforementioned facts, one of the most important concerns associated with intelligent vehicles is to secure them against malicious cyberattacks. For integrated vehicle security, fundamental defense mechanisms other than conventional authentication, firewall, or limited access of connection against malicious attacks on the internet connection have not been announced. Conventional security and privacy methods used in personal computers or cloud networks tend to be ineffective in intelligent vehicles due to their differences in vulnerability in cyberattacks.

3. BLOCKCHAIN-BASED TRUST NETWORK AMONG INTELLIGENT VEHICLES

Intelligent vehicles (IVs) are experiencing a revolutionary growth in research and industry, but still suffer from numerous security vulnerabilities. To take advantage of the opportunities of intelligent vehicles as a new platform of information technology, we need to overcome a number of challenges in implementing a secure trust network between intelligent vehicles. In modern secure communication, all data communication is encrypted and authenticated using a public/private key infrastructure with certificate-based mutual authentication between connected nodes or cloud servers. Although mutual authentication ensures that both devices can verify the authenticity of each other, these traditional security methods are still vulnerable to secure data sharing between intelligent vehicles [10–12]. A fundamental requirement of ITSs is the guarantee of provenance, immutability, and an error-resilient network between intelligent vehicles.

The characteristics are described as follows:
- Provenance: Each participant knows where the asset came from and how its ownership has changed through transactions.

- Immutability: No participant can tamper with a transaction after it has been recorded in the database (Ledgers).
- Resilience: If a transaction is in error, the corruption is readily apparent, and everyone is made aware of it. A new transaction can be used to recover the error, and the transaction is then restored. In the event that some nodes are offline or under attack, the whole network operates as usual.

Blockchain has been recognized as an emerging technology with the potential to disrupt all traditional industries. The impact of blockchain technology has been experienced in the financial sector already with cryptocurrencies like Bitcoin, Ethereum, and other crypto coins. Currently, the blockchain technology has exploited to applications such as smart contacts, medical and healthcare, and business logistics, then, attempts are being made to incorporate blockchain into more sophisticated areas such as the security of IoT. In the following section, we will discuss "How the blockchain can enhance security and privacy capabilities in a trusted environment-based intelligent vehicle communication network".

3.1 A Brief Review: How the Blockchain-Based Trust Network Works

We are currently exploring the fundamental question of "How can the trust network among anonymous participants be formed by blockchain technology?" A blockchain is a decentralized ledger that contains data of all transactions performed across a peer-to-peer network. Then, how can the concept of a shared distributed ledger make a secure and immutable record of all transactions on the network? Fig. 5 shows an overview of the basic mechanism of Blockchain.

Blockchain uses a well-known asymmetric private/public key and hash cryptography mechanism to validate the authentication of transactions. The private key is used to sign the transactions made by each node. All transactions signed with a private key are broadcast to all network nodes. All mining nodes of the network collect new transactions and works on finding a resolution for a given consensus rule. When a node finds a consensus of the given mechanism, it broadcasts the block to all nodes. All other nodes receive the block and, only if all transactions therein are valid, all nodes express their acceptance of the block by creating the next block in the chain. The first node that encapsulates the block successfully, i.e., finds a consensus resolution, obtains a reward of cryptocurrency, which may be a worthwhile goal, to motivate each mining node to elaborate to find consensus of the

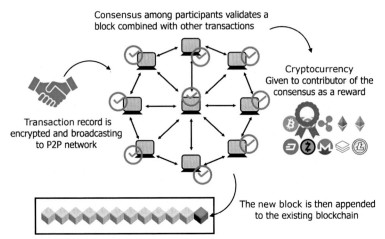

Fig. 5 Overview of the basic mechanism of blockchain.

next block. This consensus with reward and incentives system becomes the driving force for participants to maintain the blockchain without being breakdown.

The transaction data contained in these blocks become immutable, and this can be considered to be trusted and secured, because transactions are recorded by synchronized time stamps and grouped into cryptographically secured blocks that are organized in chains [13].

Depending on its openness, we classify the blockchain with public architecture like Bitcoin and Ethereum, which operate with anonymous untrusted participants with permission-less access, or with permissioned access, which deals only with approved members. In the public blockchain, each node can take participate in the consensus process, where, transactions are automatically approved and recorded by mass participation [14].

3.2 Blockchain-Based Trust Network Among Intelligent Vehicles

Blockchain is able to assure three major risks to build a trust network among intelligent vehicles without central control authority, protecting the privacy and security against cyberattacks, and guaranteeing resilience under unpredictable failure or attack on the vehicular cloud network.

Blockchain technology is based on peer-to-peer connections that allow intelligent vehicles to interact with each other without any intervention of third trusted authorities. A large number of vehicle nodes ensure the resilience of the blockchain network, and even when some nodes are unavailable due to infected nodes with malicious software or those under a cyberattack, the correct blockchain will still be accessible to vehicles. Even if some nodes are offline or under attack, blockchain can make the vehicular network operates as usual. All transactions in blockchains are time-stamped, and encrypted with private keys so vehicle users can easily trace the history of transactions and track accounts at any historical moment. This feature also allows each participant to track and trace the provenance of data or assets, i.e., the valid information regarding the source of the asset and how its ownership has changed through transactions. This tracking capability prevents potential attack by sending fraudulent data or tampered messages in V2X communications, securing the network against fraudulent security. An attacker may attempt to deanonymize the ID of the vehicle or tracking privacy information of the user, and each transaction contains privacy-sensitive data that endangers the privacy of the user. The blockchain can ensure protection of the privacy of the user by using a hash function, and encryption using asymmetric cryptography.

3.3 Use Case/Application of Blockchain for Intelligent Vehicles

In this section, we will discuss some ongoing research and the implementation of blockchain applications for intelligent vehicles.

Several research organizations and industries are working on blockchain technology implementation for the intelligent vehicles. Yuan et al. [15] have proposed the blockchain technology for ITS for the establishment of a secured, trusted, and decentralized autonomous ecosystem and proposed a seven-layer conceptual model for the blockchain. Leiding et al. [16] have also proposed the blockchain technology for the VANET. They have combined the Ethereum blockchain-based smart contracts system with the VANET. They have proposed a combination of two applications for vehicles, namely, mandatory applications (traffic regulation, vehicle tax, vehicle insurance) and optional applications (applications that provide information and updates on traffic jams and weather forecasts) of vehicles. They attempted to connect the blockchain with VANET services. Blockchain can use multiple other functionalities such as communication between vehicles, provide security, and provide peer-to-peer communication without disclosing personal information. Dorri et al. [17] proposed the blockchain technology mechanism without disclosing any private information of vehicles users to provide and update the wireless remote software and other emerging vehicles services. Rowan et al. [18] described the blockchain technology for securing intelligent vehicles communication through the visible light and acoustic side-channels. They have verified their proposed mechanism through a new session cryptographic key, leveraging both side-channels and blockchain public key infrastructure. However, thus far, the focus has been on services and their business models, and safe and secure trust environments for the intelligent vehicle communication has not been discussed.

In addition, Toyota has declared that they are working on blockchain technology for the connected vehicle project [6,19]. Toyota research institute (TRI) is collaborating with academic institutes and industry for this blockchain for intelligent vehicles projects. They believe that blockchain technology has emerging features for the storage and use of driving data of automated vehicles. TRI and its collaborators are concentrating on data sharing, ride sharing and transactions, and user-based insurance. ZF motion and mobility company [20] is working with the UBS Swiss bank on blockchain technology for electric cars. They have proposed the mobile e-wallet concept for secure vehicle-related payments such as toll payment, energy charge payment, and car sharing/ride sharing.

3.3.1 Toyota's Blockchain Projects

As mentioned, the automotive industry is the next target of blockchain technology. TRI has already started three blockchain projects on autonomous technology [6,19]. These projects are as follows:

- *Driving/testing data sharing:* Intelligent vehicles connected to the cloud are aware of their environment through onboard sensors, all of which generate valuable driving data. Blockchain technology can provide a decentralized platform of frameworks, which allows creators and perspective customers to share and monetize their driving information in a secure marketplace. Toyota aims to development of a blockchain framework to share driving data while preserving ownership of the data. The goal of this attempt is similar to the approach of the blockchain initiative of the open music initiative, which creates digital property rights in the music industry [21].

- *Car/rideshare transactions:* Blockchain enables storage of the vehicle usage data and information about users (owner, passengers, driver, etc.). Toyota is attempting to build a blockchain-based tool to empower vehicle owners to monetize their asset by sharing their car by selling rides or cargo space. This is a service based on a smart contract between two parties, and blockchain helps to manage the financial transaction between them without any third party intervention, thereby saving transaction surcharges.

The system will also provide connectivity to vehicle functions for remote access to locking/unlocking doors and on/off control of the engine.

- *Usage-based insurance:* By allowing collection of driving data and storage in a blockchain, vehicle owner's pay lower insurance costs by providing their driving data for evaluation of the driving style to influence safe driving. The insurance company can correctly evaluate the safety of the driving habits from the driving data stored in blockchain.

Blockchains and distributed ledgers may enable the pooling data from vehicle owners, fleet managers, and manufacturers to shorten the time to reach this goal, thereby advancing the safety, efficiency, and convenience benefits of autonomous driving technology. TRI insists that hundreds of billions of miles of human driving data may be needed to develop safe and reliable autonomous vehicles. Through an open-source approach to software tools, TRI is creating a user consortium and hopes to stimulate a rapid adoption of blockchain by other companies developing autonomous vehicles and providing mobility services. TRI is inviting current and future partners to collaborate on further development of blockchain and distributed ledger technology applications in vehicle data and services.

TRI is working on these projects with MIT Media Lab and several industry partners. These partners are BigchainDB (a German company), which is building the driving/testing database; Oaken Innovation (Canadian Company) developing peer-to-peer car sharing and mobility tokens based payments system; Commuterz, an Israeli Startup Company, developing a carpooling solution; Gem, (a US-based company) working with Toyota Insurance Solutions and anther Japanese company Insurance on a usage-based insurance system.

3.3.2 Blockchain-Based ITSs

The China Academy of Sciences proposed an ITS-oriented seven layers stacked open system model of blockchain similar to the well-known open system interconnection (OSI) reference model which defines a networking framework to implement protocols in seven layers as shown in Fig. 6 [15]. It provides a conceptual framework to understand complex relationships of layer to layer interactions that are occurring at each a vertical stack and internetworks. Although its usefulness for the reference model to see how the system can work over the network, OSI has been rarely actually implemented. The proposed seven-layer blockchain model provides a reference for characterizing and standardizing the typical architecture and major components of blockchain systems, and briefly describes its underlying key techniques.

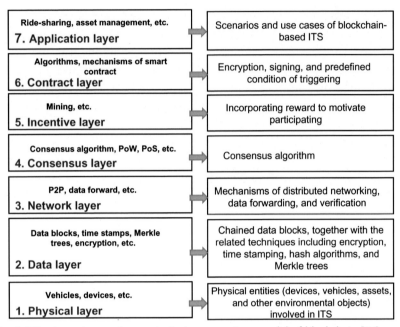

Fig. 6 ITS-oriented seven layers stacked open system model of blockchain [15].

3.3.3 CUBE—Autonomous Car Network Security Platform Based on Blockchain

CUBE aims to provide security platform for autonomous vehicles based on blockchain technology as a solution to autonomous vehicles' security. Autonomous vehicles rely heavily on over-the-air wireless technology, which increases the risk of malicious attacks on autonomous vehicles [22]. CUBE focuses on a tool to secure over-the-air software updates or contents exchange. CUBE has developed over-the-air technology that enables remote software diagnostics, and installation and bug patches of existing automotive software based on blockchains of the Ethereum platform. The CUBE token links every blockchain-based service such as over-the-air services, driving-related data sharing, and insurance services.

An example of a CUBE token application is the linking of driving-related information from vehicles to data consumers who need driving information. The vehicle owners generate the driving information while driving the car, receiving CUBE tokens in return. CUBE digitizes this information and links it into a blockchain. Data consumers who need this information provide the services in exchange for CUBE tokens. When an automobile generates information, all nodes send, utilize, and receive it. Automobiles exchange their data as nodes, and the data consumers who need this information provide various services and receive tokens. Information consumers such as automobile and insurance companies can pay providers for traffic information using CUBE tokens. CUBE is an imminent technology for commercialization; it applies blockchain to intelligent vehicle security and has a payment and reward system using cryptocurrency. Using a CUBE token, one can receive services at CUBE affiliates such as gas stations, car dealers, repair shops, and insurance companies, and affiliate companies can provide services such as discounts, vehicle maintenance, vehicle purchase discounts, insurance benefits, and cash-back services. CUBE announced that its autonomous vehicle security platform technology based on blockchain will be released in 2019. The key of the autonomous vehicle security platform is that technology ensures trust using blockchain. CUBE uses blockchain technology to ensure the security of autonomous wireless networks. CUBEs algorithm adopts the blockchain technology proposed by Dorri et al. [17], which is a multisig mechanism, without disclosing any private information of vehicles users to provide and update the wireless remote software and other emerging vehicles services.

3.3.4 Use Case of Solving Intersection Deadlock Problem for Autonomous Vehicles

Deadlock is the state in which no operation can progress. Autonomous vehicles negotiate the right-of-way before traversing each intersection without traffic signals (on a first-come first-served basis), because the car reaching the intersection first has passing priority. If four autonomous vehicles arrive at a four-way intersection almost simultaneously, none can proceed until determination of the priority, which results in the deadlock at intersections without traffic signals. A cooperative traffic management system may allow high performance at intersections; however, it may cause considerable maintenance costs or inefficiency because most intersections without traffic signals are placed in low-traffic regions. Kim and Singh proposed a trust point-based blockchain technology for intelligent vehicles, as well as a solution to the deadlock scenario [23]. As shown in Fig. 7, four vehicles, namely, IV-1, IV-2, IV-3, and IV-4 approach the intersection. When the vehicles are about to reach the intersection, they broadcast their signals. A consensus is made regarding the vehicle the earliest timestamp associated with the block and which should move first. Then, the same procedure is followed for the other three (possibly more) vehicles. Fig. 7 shows an example of an intersection scenario for IVs, where the arrival times of IV-1, IV-2, IV-3, and IV-4 at the intersection are 10:06, 10:02, 10:04, and 10:03, respectively. IV-2 crosses the intersection first, after consensus is reached.

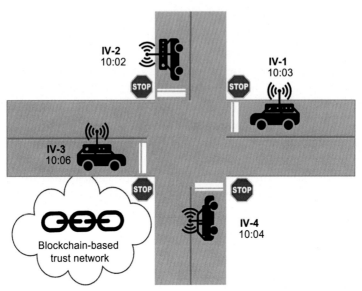

Fig. 7 Example of deadlock at traffic intersection without signals (on a first-come first-served basis) for autonomous vehicles.

4. CHALLENGES AND FUTURE RESEARCH DIRECTIONS OF BLOCKCHAIN IN INTELLIGENT VEHICLES

Blockchain technology is gaining momentum and creating new business opportunities both in the automotive and ICT industries to untangle the potential cybersecurity threat of intelligent connected vehicles; however, some challenges lie ahead.

While many applications are being experimented with in various fields, there remain some unsolved issues that prevent us from stating that blockchain technology has perfect integrity under any environmental condition of implementation.

The main challenge of blockchain consensus protocols such as Bitcoin, Ethereum, Ripple, and Tendermint is that every fully participating node in the network must process every transaction. Recall that every single node on the network processes every transaction and maintains a copy of the entire state. While a decentralization consensus mechanism offers some critical benefits, such as fault tolerance, and a strong guarantee of security, political neutrality, and authenticity, it comes at the cost of scalability.

4.1 Challenging Issue of Blockchain for Intelligent Vehicles

Blockchain technology is a breakthrough in cybersecurity, as it can ensure the highest level of data confidentiality, availability, and security. However, the complexity of the technology may cause difficulties with development and real-world use.

In order to bring the blockchain to intelligent connected vehicles, participants of the vehicle cloud need to overcome a number of technical challenges due to the limited capability of in-vehicle power and storage, and many other factors. Five main challenges are identified and explained as follows:

(a) *Scalability*: Scalability means the capability of the blockchain system to cope with and handle a growing amount of distributed ledger database. Scaling the blockchain is a known challenge and has been an active area of research for several years. Scalability is a primary concern of blockchain architecture for intelligent vehicles, because each vehicle has a limited amount of storage capacity and power budget to work for a mining node or distributed ledger. In the traditional blockchain mechanism, each node consumes a significant power for mining, and a data storage server is required to permanently store shared data of the distributed ledger, which is almost unable to implement in vehicle environment. One potential resolution is periodic backup of the in-vehicle data to external backup storage [17]. However, managing the ever-growing data size of the shared ledger may be burden in practical use cases.

Another solution for keeping the network lighter is the use of a decentralized storage service such as Swarm. Swarm is a peer-to-peer file sharing protocol for Ethereum that lets you store application code and data off the main blockchain (MB) in swarm nodes, which are connected to the Ethereum blockchain, and later exchange this data on the blockchain. The basic premise here is that instead of nodes storing everything on the blockchain, they only store data that is more frequently requested locally and leave other data on the "cloud" via Swarm.

As a result, all public blockchain consensus protocols that operate in such a decentralized manner make the tradeoff between low transaction throughput and high degree of centralization. In other words, as the size of the blockchain increases, the requirements for storage, bandwidth, and computing power required by fully participating in the network increases. At some point, it becomes sufficiently unwieldy that it is only feasible for a few nodes to process a block—leading to the risk of centralization. To scale, the blockchain protocol must develop a mechanism to limit the number of participating nodes needed to validate each transaction, without losing the network's trust that each transaction is valid. It may sound simple, but is very difficult in a technological point of view.

Kim et al. proposed a branch-based blockchain composed of a local dynamic blockchain (LDB) and a MB enabled with intelligent vehicle trust point [24]. Fig. 8 shows the blockchain technology with Local Dynamic Block and Main Block for V2X communication proposed by Kim et al. LDB with circular queue architecture stores only local temporal data shared in local area. If data size exceeds the memory capacity of LDB, new incoming data are overwritten. If LDB receives any unusual event data (not a regular event), then it forwards it to MB for permanent storage. LDB can branch or merge depending on the amount of data traffic and the workload required to handle in real time. Fig. 9 shows branching and merging operation of a local dynamic blockchain. If a LDB branches, original LDB is divided into two child branches. Further, if two branches are merged, both parent branches must contain a block on the latest level.

(b) *Transaction speed and computing power:* To add a new transaction, consensus of among network participants must be recorded in the distributed ledger of the blockchain. However, the implemented blockchain consensus mechanism does not yet offer significant improvements to

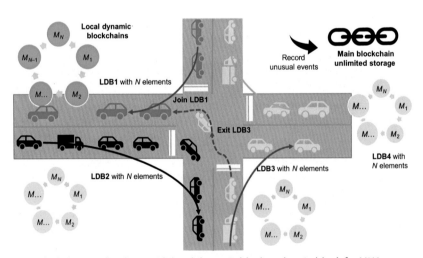

Fig. 8 Blockchain technology with local dynamic block and main block for V2X communication [24].

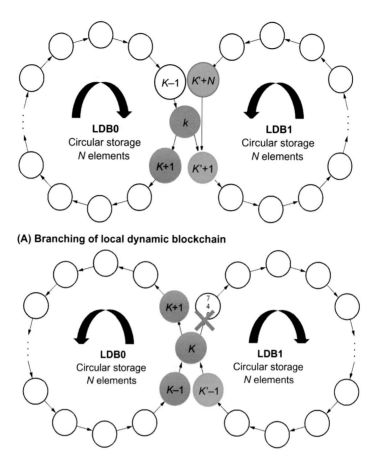

(A) Branching of local dynamic blockchain

Fig. 9 Branching and merging operation of a local dynamic blockchain, where LDB0 splits at the *k*th element of the block to LDB1.

transaction speeds. In the case of commonly used algorithms, the period of consensus to encapsulate new block is 10 min for proof-of-work (for Bitcoin) and 17 s for proof-of-stake (for Ethereum). In contrast, the average confirmation time of Bitcoin is approximately 250 min; the longest time ever recorded, in February 2018, was more than 2500 min (approximately 42 h). These existing consensus algorithms of the blockchain are too slow for intelligent vehicles to record messages between vehicles in real time. Moreover, mining nodes consume considerable computing power. To resolve the barrier of speed and power budget, some new, innovative approaches have to be introduced to improve these essential blockchain characteristics.

An Ethereum node's maximum theoretical transaction processing capacity is over 1000 transactions per second. Unfortunately, this is not the actual throughput owing to Ethereum's "gas limit," which is currently approximately 6.7 million gas on average for each block [25]. Hence, the "gas limit" for each block determines how many transactions will fit in a block based on the gas limit specified by each transaction in the block. Similarly, Bitcoin, despite having a theoretical limit of 4000 transactions per second, currently has a hard cap of approximately seven transactions per second for small transactions and three per second for more complex transactions. Unfortunately, none of the solutions perfectly address the transaction speed and computing power problems.

Both of these solutions aim to solve the Bitcoin-specific issue where the Bitcoin blockchain has a built-in hard limit of 1 MB per block, which caps the number of transactions that can be added to a block. As a result, Bitcoin has been facing delays (sometimes hours and even days) in processing and confirming transactions for a while now. Similarly, as noted in the previous section, Ethereum also faces limitations in its ability to scale.

(c) *Confidentiality vs Transparency:* Despite great features of the blockchain technology, many regulators are being challenged by the type of data that should reside within the distributed ledge. Many people feel that encrypted information managed with public and private keys should support the confidential information being stored on the blockchain. In the decentralized model, all parties share information. No private contractual data should be stored on a distributed ledger, encrypted, or otherwise. Shared ledgers should contain the bare minimum information, interpretable only by those with a need and right to know, to permit notification, synchronization, and confirmation. For confidentiality, confidential and sensitive data should reside on a blockchain as it provides an extra layer of security with immutable records.

(d) *Privacy:* From a security perspective, researchers developed various techniques targeting privacy concerns focused on personal data. Data anonymization methods attempt to protect personally identifiable information. Recent research has demonstrated how anonymized datasets employing these techniques can be deanonymized [2], given even a small amount of data points or high dimensionality data.

Although we often use the terms "confidentiality" and "privacy" interchangeably in our daily lives, they are distinct terms from a legal standpoint. We generally use the term confidentiality in the context of protecting data (e.g., transaction details, account and wallet balances, and contents of a contract) from unauthorized third parties. We use the term privacy to refer to protection from intrusion into the identity of blockchain participants and parties to transactions.

- Data Ownership: It focuses on ensuring that users own and control their personal data. As such, the system recognizes the users as the owners of the data and the services as guests with delegated permissions.
- Data Transparency and Auditability: Each user has complete transparency over the data collected and how they are accessed.

(e) *Locality of ground transportation/cyberspace is globally synchronized:* Seamless integration of local vehicular networks and globally distributed storage platform pose numerous challenges since all these building blocks are heterogeneous in terms of their dependencies on infrastructure and software elements. Ground transportation is localized, but blockchain is a global cyberspace. Local transactions do not need to be shared on the globally distributed ledger, which may reduce unnecessary data traffic. The problem of locality and globally synchronized ledger can be solved by focusing on data centric aspects of vehicular communication networks. The detection of the abnormality of cyberattacks for other regions by proof of driving is a potential solution.

Although blockchain technology can be applied to almost any business, intelligent vehicle industry may face difficulties integrating it. It is quite challenging to employ this technology in vehicle cloud systems, for instance, as it may be time-consuming to replicate vehicle network as blockchains and refine them. Blockchain applications can also require the complete replacement of existing systems; thus, business parties should consider this before implementing blockchain technology.

KEY TERMINOLOGY AND DEFINITIONS

Consensus in Blockchain It means decision-making among participant nodes: For a transaction to be valid, all participants must agree on the validity of a predefined algorithm or rule, such as proof-of-work, proof-of-stake, proof-of-concept, or else. The consensus involves conducting two key crucial functions of the blockchain technology. First, consensus protocols allow newly generated data block to be appended to a distributed Ledger, while ensuring that every block in the chain is truly valid and keeping participants incentivized to do mining. It prevents malicious hackers controlling or breaking down the blockchain system. Second, the consensus rule guarantees that a single blockchain keeps growing and is followed.

Cybersecurity Cyber security is defined as the protection of systems, networks, and data in cyberspace and is a critical issue for all businesses. Cybersecurity, computer security, or IT security is the protection of computer systems from the theft and damage to their hardware, software, or information, as well as from disruption or misdirection of the services they provide. Cybersecurity includes controlling physical access to the hardware, as well as protecting against harm that may be caused as a result of network access, data, and code injection. In the vehicle networks, safety, security, and privacy are key issues of the protection of computerized networks from the theft and damage to their hardware, software, or information, as well as from disruption or misdirection of the services they provide.

Intelligent and Connected Vehicles Vehicles which are controlled by a complex computer with connectivity of both the in-vehicle network and wireless communication network. Autonomous vehicle is a state-of-the-art of system where actuation and decision-making maneuvers operate concurrently with coping with environmental information coming from sensors and communication networks during driving. The intelligent vehicles will be connected to almost everything, they have become a part of the internet of things.

Scalability Scalability is the capability of a system, network, or process to handle an increasing amount of work, or its potential to be enlarged to accommodate this growth. Scalability is a characteristic of a system, model, or function that describes its capability to cope and perform under an increased or expanding workload. A system that scales well is able to maintain or even increase its level of performance or efficiency when tested by larger operational demands.

Trust Network Among Intelligent Vehicles This is a networking environment that is secured against malicious cyberattacks on V2V, V2X communication, and vehicular cloud networks.

REFERENCES

[1] E.B. Hamida, K.L. Brousmiche, H. Levard, E. Thea, in: Blockchain for enterprise: overview, opportunities and challenges, International Conference on Wireless and Mobile Communications, 2017.

[2] G. Zyskind, O. Nathan, in: Decentralizing privacy: using blockchain to protect personal data, IEEE Security and Privacy Workshops, 2015, pp. 180–184.

[3] Z. Zheng, S. Xie, H.N. Dai, H. Wang, Blockchain challenges and opportunities: a survey, Int. J. Web Grid Serv. (2016).

[4] N. Kshetri, Blockchain's roles in strengthening cybersecurity and protecting privacy, Telecommun. Policy 41 (2017) 1027–1038.

[5] C. Mauro, C. Lal, S. Ruj, A Survey on Security and Privacy Issues of Bitcoin, arXiv:1706.00916, 2017.

[6] A.G. Simpson "Toyota, MIT Lab Eye Using Blockchain in Insurance Rating of Driverless and Shared Vehicles" https://www.insurancejournal.com/author/andrew-simpson, last visited March 2018.

[7] A. Papathanassiou, A. Khoryaev, Cellular V2X as the essential enabler of superior global connected transportation services, IEEE 5G Tech Focus 1 (2) (2017).

[8] Connected Vehicle Cloud—Under the Road, Ericsson, www.ericsson.com.

[9] S.K. Datta, R.P.F. Da Costa, J. Härri, C. Bonnet, in: Integrating connected vehicles in internet of things ecosystems: challenges and solutions, IEEE International Symposium on World of Wireless, Mobile and Multimedia Networks, 2016, pp. 1–6.

[10] M. Singh, S. Kim, Safety Requirement Specifications for Connected Vehicles, arXiv:1707.08715, 2017.

[11] J.H. Park, J.H. Park, Blockchain security in cloud computing: use cases, challenges, and solutions, Symmetry 9 (8) (2017) 164.

[12] P.K. Sharma, S.Y. Moon, J.H. Park, Block-VN: a distributed blockchain based vehicular network architecture in smart city, J. Inf. Process. Syst. 13 (1) (2017) 184–195.

[13] C. Cachin, M. Vukolic, Blockchain Consensus Protocols in the Wild, arXiv:1707. 01873, 2017.

[14] C. Xu, K. Wang, M. Guo, Intelligent resource management in blockchain-based cloud datacenters, IEEE Cloud Comput. 4 (6) (2017) 50–59.

[15] Y. Yuan, F. Wang, in: Towards blockchain-based intelligent transportation systems, IEEE International Conference on Intelligent Transportation Systems (ITSC), 2016, pp. 2663–2668.

[16] B. Leiding, P. Memarmoshrefi, D. Hogrefe, in: Self-managed and blockchain-based vehicular ad-hoc networks, Proceedings of the ACM International Joint Conference on Pervasive and Ubiquitous Computing, 2016, pp. 137–140.

[17] A. Dorri, M. Steger, S.S. Kanhere, R. Jurdak, Blockchain: a distributed solution to automotive security and privacy, IEEE Commun. Mag. 55 (12) (2017) 119–125.

[18] S. Rowan, M. Clear, M. Gerla, M. Huggard, C. Mc Goldrick, Securing Vehicle to Vehicle Communications Using Blockchain Through Visible Light and Acoustic Side-Channels, arXiv:1704.02553, 2017.

[19] "Toyota's Vision of How Blockchain Will Change the Auto Industry", http://www. thebanksreport.com, last visited March 2018.

[20] Jen Clark, "ZF, UBS and IBM Bring Blockchain to In-Vehicle Payments", www.ibm. com/blogs/internet-of-things/zf-ubs-ibm-vehicle-payments/ last visited March (2018).

[21] R. Xu, L. Zhang, H. Zhao, Y. Peng, in: Design of network media's digital rights management scheme based on blockchain technology, IEEE International Symposium on Autonomous Decentralized Systems, 2017, pp. 128–133.

[22] CUBE, Autonomous Car Network Security Platform Based on Blockchain, White Paper, Cube, 2017.

[23] M. Singh, S. Kim, Intelligent Vehicle-Trust Point: Reward Based Intelligent Vehicle Communication Using Blockchain, arXiv:1707.07442, 2017.

[24] M. Singh, S. Kim, Blockchain Based Intelligent Vehicle Data Sharing Framework, arXiv:1708.09721, 2017.

[25] K. Croman, C. Decker, I. Eyal, A.E. Gencer, A. Juels, A. Kosba, A. Miller, P. Saxena, E. Shi, E.G. Sirer, D. Song, R. Wattenhofer, On scaling decentralized blockchains, in: International Conference on Financial Cryptography and Data Security (FC 2016), Lecture Notes in Computer Science, 9604, 2016, pp. 106–125.

FURTHER READING/REFERENCES FOR ADVANCE

[26] A. Lei, H. Cruickshank, Y. Cao, P. Asuquo, C.P.A. Ogah, Z. Sun, Blockchain-based dynamic key management for heterogeneous intelligent transportation systems, IEEE Internet Things J. 4 (6) (2017) 1832–1843.

[27] M. Gupta, Blockchain for Dummies, John Wiley & Sons, 2017.

[28] P. Fairley, Special report: blockchain world, IEEE Spectr. (2017) 24–56.

[29] M. Niforos, V. Ramachandran, T. Rehermann, BLOCKCHAIN-Opportunities for Private Enterprises in Emerging Markets, IFC (International Finance Corporation), 2017.

[30] J. Sun, J. Yan, K. Zhang, Blockchain-based sharing services: what blockchain technology can contribute to smart cities, Financ. Innov. 2 (2016) 26.

[31] Consensus—Immutable Agreement for the Internet of Value, kpmg.com/socialmedia, 2016.

[32] K. Toyoda, P. Mathiopoulos, I. Sasase, T. Ohtsuki, A novel blockchain-based product ownership management system (POMS) for anti-counterfeits in the post supply chain, IEEE Access 5 (2017) 17465–17477.

[33] T.D. Smith, in: The blockchain litmus test, IEEE International Conference on Big Data (Big Data), 2017, pp. 2299–2308.
[34] D. Yang, J. Gavigan, Z. Wilcox-O'Hearn, Survey of Confidentiality and Privacy Preserving Technologies for Blockchains, R3 Research Report, 2016.
[35] J.M. Easton, Blockchains: a distributed data ledger for the rail industry, in: Innovative Applications of Big Data in the Railway Industry, IGI Global, 2017, ISBN13: 9781522531760, pp 27–39.
[36] M. Cebe, E. Erdin, K. Akkaya, H. Aksu, S. Uluagac, Block4Forensic: An Integrated Lightweight Blockchain Framework for Forensics Applications of Connected Vehicles, arXiv:1802.00561, 2018.

ABOUT THE AUTHOR

Shiho Kim is a professor in the school of integrated technology at Yonsei University, Seoul, Korea. His previous assignments include, being a System on chip design engineer, at LG Semicon Ltd. (currently SK Hynix), Korea, Seoul [1995–1996], Director of RAVERS (Research center for Advanced hybrid electric Vehicle Energy Recovery System), a government-supported IT research center. Associate Director of the Yonsei Institute of Convergence Technology (YICT) performing Korean National ICT consilience program, which is a Korea National program for cultivating talented engineers in the field of information and communication Technology, Korea [2011 – 2012], Director of Seamless Transportation Lab, at Yonsei University, Korea [since 2011 to present].

His main research interest includes the development of software and hardware technologies for intelligent vehicles, blockchain technology for intelligent transportation systems, and reinforcement learning for autonomous vehicles. He is a member of the editorial board and reviewer for various Journals and International conferences. So far he has organized two International Conference as Technical Chair/General Chair. He is a member of IEIE (Institute of Electronics and Information Engineers of Korea), KSAE (Korean Society of Automotive Engineers), vice president of KINGC (Korean Institute of Next Generation Computing), and a senior member of IEEE. He is the coauthor for over 100 papers and holding more than 50 patents in the field of information and communication technology.

> CHAPTER THREE

Blockchain Technology: Supply Chain Insights from ERP

Arnab Banerjee

Principal Consultant, Infosys Limited, India

Contents

Abstract

The chapter provides a high level understanding of how ERP system alongside Blockchain technology will be a powerful tool to improve supply chain operations. The chapter details out how the two technologies will complement each other in every aspect of supply chain functions bringing in transparency, efficiency, and cost reduction. The chapter considers every aspects of supply chain for an ERP enabled organizations and details out use cases for master data, engineering design, sales process, procurement process, demand and supply planning process, manufacturing process, and logistics management processes. The chapter provides use case details and high level

Advances in Computers, Volume 111
ISSN 0065-2458
https://doi.org/10.1016/bs.adcom.2018.03.007

understanding of technology for product provenance and how it can bring in supply chain transparency using blockchain. The chapter illustrates theoretical and conceptual model for use of open and permissioned blockchain in different supply chain applications with real life practical use cases as is being developed and deployed in various industries and business functions. The chapter also emphasizes the use of blockchain in distribution industry and how it can solve pertinent problems as it exists today in the distribution supply chain. The chapter ends with an outlook of blockchain how it will shape the future to come and challenges which lies there within.

1. A BRIEF INTRODUCTION TO SUPPLY CHAIN, ERP, AND IMPLICATION OF ERP ON THE SUPPLY CHAIN

Supply chain management (SCM) is an established and essential business process in every organization today. SCM uses specific processes to connect a customer's requirement with a supplier through a channel. Supply chain practices involve numerous subprocesses. Over the years, technology developments have effected significant change and improvement in these practices across demand, operations, supply, and distribution. There has been tremendous development in technology to rationalize and improve supply chain processes. For instance, in the last four decades, supply chain technology has evolved from material requirement planning (MRP) to manufacturing resource planning (MRP II), to enterprise resource planning (ERP), and, finally, to advanced supply chain planning and optimization (APS/APO). All these systems depend heavily on technology. Of all the technologies, ERP is having the most noticeable impact on the supply chain improving its operations and maturing its practitioners. There has been numerous research findings which supports that ERP has positively impacted a firms supply chain performance [1–3]. Today, almost all leading companies run computerized ERP and supply chain management software (Brody, 2017).

Over the past two decades, ERP has had the most significant impact on supply chain business processes. However, despite employing large-scale digital infrastructure, most companies struggle with visibility into demand, orders, and supply. Further, they lack insights into where their products are at any given moment outside their organizations network. Li et al. [4] and Ince et al. [1] acknowledge that ERP improves supply chain effectiveness and increases its performance and competitive advantage. Koh et al. [35] highlighted the need of ERP II to share information across different ERPs in supply chain. The concept of ERP II clearly states that there is gap existing between the ERP systems (across organizations) resulting in fragmented information. The research of Koh el al. [36], Addo-Tenkorang and Helo [37] highlights the need for an open system to make information available to supply chain trading partners. The reason for this gap is ERP primarily improves internal organization operations but hardly impacts anything outside the organizations. Thus, as a software product, ERP continues to evolve, particularly with cloud technologies and analytics but still lacks connectivity across its stakeholder outside the organizations.

ERP can be described as a system that integrates an organization and its resources to maximize resource utilization and improve planning. ERP transforms how the organization interacts internally as well as externally. It provides a common platform for different organizational departments like production planning, material planning, operations, inventory, warehousing, design, sales, distribution, accounting, and customer service to act as a unified entity. This enhances the customer experience, operating efficiency, delivery execution, and help track and improve supplier performance. ERP seamlessly marries information flow to material flow creating a homogeneous environment for computerized business processes. It removes physical boundaries, strengthens centralized processes, and globalizes IT architecture.

ERP systems are generally sold as packaged software by companies like Oracle, SAP, Infor, etc. ERP systems can be standard or bespoke. An important point to note is that while ERP transforms an organization's internal operations and the way it interacts with stakeholders, it does not connect one organization to another or create a network among partners/stakeholders. Even today, most companies rely on point-to-point messaging systems like XML or EDI to exchange data and in more recent years SOA architecture. So what is missing in today's IT supply chain world is a network that can connect ERP systems irrespective of their original developers and underlying technology. This is where blockchain can play a pivotal role.

With blockchain, ERP systems can come together to form an integrated platform across partners/stakeholders, providing data that cannot be manipulated (true record) and an audit trail for transactions that happen within the extended ERP network. This is the key tenet of this chapter. The focus of this chapter is to connect the supply chain processes in ERP system across industries and domains with blockchain in order to help practitioners achieve greater efficiency and cost benefit in final delivery of product.

2. ERP TRANSACTIONS, INDUSTRY USE CASES, AND GENERIC USE CASES WITH BLOCKCHAIN

As mentioned earlier, ERP transforms internal supply chain operations and processes from order booking to order fulfillment. Supply chain interacts across partners like suppliers, manufacturers, service providers, customers and so does their ERP systems. All these processes, subprocesses, and ERP systems can be integrated with a centralized blockchain for further improvement beyond what is possible with ERP. But first, companies must develop key integration capabilities so that blockchain can integrate with enterprise systems. The underlying principle is to connect all the partner ERP systems to a blockchain network. This can be an open network or a private permissioned network. This combined entity of ERP and Blockchain can complement each other's capability and provide a seamless supply chain transaction across partners.

In open architecture, all partners (ERP systems) can join/connect to the network. Here, miners play a significant role. Transactions as passed by members to the system and are verified and authenticated by miners before they are posted as a block. Organizations can use product, industry, or business-specific blockchains that are permission controlled and leverage a private permissioned blockchain network. In a private blockchain network, the ERP systems are connected to the blockchain. Smart contracts play a vital role in permissioned blockchains. Here, member entry can be decided either by existing members or by the network owners. Smart contracts determine the underlying rules of processes, validations, and interactions.

In either case, Blockchain with its distributed ledger will act as the central backbone for all its partner ERP systems or as a centralized repository across all partners. The data originating from ERP transactions can be channelized to blockchain as per the need, design, or smart contract of the use case. This extended enterprise platform can also drive multiple use cases by deriving value from all the connections across the partner ERP systems. These connections will help commercial operations/transactions, compliance, service operations efficiency, extended supply chain planning capabilities, and data consistency across enterprises, product provenance, and data sharing abilities, thereby enabling informed decision-making.

There are many industry-specific consortiums that bring together data, suppliers, customers, governments, supply chains, and even individuals onto a single platform to achieve a common goal. Most of them are privately owned like Trade Service, OceanConnect, Exostar, and SiliconExpert. Companies subscribe to these consortium for data feeds. These industry consortium get the feed from manufacturers, etc. and enrich those data and feed it to their customers. The customers who are also a member of the consortium have their own transactional system. As the feeds are from multiple manufacturer or partners and that data is fed into many member systems the data lacks standardization. This also leads to lack of trust and independence. But, with Blockchain technology, industry-specific blockchain consortiums can maintain standardization, transparency, objectivity, and neutrality. Blockchain can bridge gaps in technology, transparency, and concurrency among various ERP systems so they can integrate effectively. This improves supplier relations, customer management, and margins while delivering cost savings and value generation on end products savings for all members.

The future sections of the chapter covers the various functions of ERP for supply chain and how it can work with blockchain as an integrated system.

3. MASTER DATA MANAGEMENT ENABLED BY BLOCKCHAIN

ERP systems (or any information system) transacts, processes, and makes decision based on data. The performance of enterprise systems depends heavily on the quality of data [5]. Thus, quality of the master data information should be up-to-date, accurate, and complete. Analyst firm Intelligent Business Strategies estimates that poor quality customer data costs US businesses around US $611 billion a year [6]. In reality, most data resides in private silos within a company and is not widely available. Even governmental data, which everyone has a right to view and use, is not easily available. While there have been attempts to release governmental data to consumers like citizens, businesses, and other organizations so they can make informed decisions [7], most of the data is not open. Similar is the case in business scenario and only some data is shared across consortiums like Trade Service, Industry Data Warehouse, Silicon Experts for distributors from manufacturer.

From a supply chain perspective, all enterprise systems have three sets of master data. The first set is the item (product) information as stored in the enterprise. These items also has the customer item name cross reference and the supplier item cross reference maintained. The customer item means in the customer enterprise system the same item is known by a different item name/designator and supplier item means in the supplier enterprise system the same item is known by a different item name/designator. The second set contains the customer name (business entity name) with whom the enterprise does the transactions with, stored in the enterprise as the customer master data. The third set is the supplier name (business entity name) with whom the enterprise does the transactions with, stored in the enterprise as the supplier master data.

Say a distributor or retailer sells a product to its customer. In the distributor enterprise system, the product is known as product X; in the customer's enterprise system, it is called product Y; and in the supplier's enterprise system, it is called product Z. All these names/item designator refer to the same product. Every supply chain transaction that happens across enterprises must carry these cross references. Currently, all leading ERP systems (Oracle, JD Edwards, SAP) have the functionality to maintain customer item and supplier item cross references with its own enterprise's naming convention. So, the data exists independently in every ERP system of the enterprise. This approach has several challenges in the areas of:

1. **Data setup:** The data must be set up as cross references. Further, since access to data is privileged and is not openly available, it must be provided by the business entity for setup. This is a time-consuming and effort-laden process.
2. **Data maintenance:** The data must be maintained well in the enterprise system. Changes must be updated regularly to prevent asynchrony as out-of-sync data can negatively impact business transactions.
3. **Data standardization:** Since data is independently set up in every enterprise, it lacks standardization. This leads to maintenance issues, process slowdown and transaction failure. Unstructured and inconsistent data can negatively impact business transactions.
4. **Multisystem architecture:** The data depends on enterprise systems and this architecture must be integrated across multiple systems (satellite systems), which is challenging and tedious.

These challenges highlight the need for centralized system which can act as the universal data repository. Blockchain with its capability to interact with multiple enterprises can act as centralized repository. Blockchain will master and govern the data which will be accessible to all its nodes. This will enable the centralized availability of truth (data) with all its details like characteristics, description, technical attributes, business names, etc. Blockchain being the unbreakable and impenetrable single source of truth the data will be securely available to all its nodes and any changes will be reflecting almost real time. There will be no scope of counterfeiting and other malicious attempts to jeopardize the data which is centrally available for all its members. When manufacturers, distributors, and retailers become part of the blockchain network, it will be the nodes and can streamline data management and transfer. For instance, the manufacturer can create, name, and characterize an item using a part number ID. This information is passed into the Blockchain and becomes the Blockchain part number ID. In this way, blockchain serves as the central information repository of the item along with all its technical information, and all the partners will have access to this information.

Now, when all partner ERP systems are connected to the centralized blockchain, the unique blockchain part number ID can be easily accessed by all the partners and converted to suit their systems. Any partner intending to do business can transfer the data from the blockchain and convert it for their ERP system. Thus, the blockchain also serves as a product marketplace whereby any partner wanting access to item data can refer to the blockchain and reach the partner/source. In this way, every unique part generated by the manufacturer can be associated with every partner, eliminating the need for separate customer and supplier part numbers. This will facilitate smoother transactions, lower transaction failure rates, and provide highly standardized data across all supply chain partners. This will also reduce the burden of generating Point Of Sale reports as the items designator will be same. This will also smoothen the claims and returns because of same item. This will also reduce time for product recall.

Another important type of master data is customer and supplier data also known as trading partner data. There is no standardization of customer data (name, address) and supplier data (name, address). This means that every system and company has a different way of storing this data. Currently, most companies use the Data Universal Numbering System (DUNS) to standardize and identify businesses. DUNS is a proprietary system developed and regulated by Dun & Bradstreet, Inc. (D&B) that assigns a unique numeric identifier, called the "DUNS number," to a single business entity or its legal address.

When it comes to customer and supplier data, the DUNS number can uniquely identify the entities with an address. The information is limited to the DUNS number. D&B is currently testing a project that allows its customers to verify the identity of a potential business partner through a unique

Blockchain identifying number that corresponds to the traditional D&B DUNS number [8].

Here is how DUNS will use blockchain. Say Company A in the United States wants to buy a product from Company B in Singapore. As these two firms have never done business before, A wants to verify the authenticity of B. Since A is part of the D&B blockchain consortium, it can download a node from the cloud where this blockchain instance resides. Typically, each node in the blockchain will be automatically updated with all the data contained as the blocks (from various companies, like B) are validated and updated with transactions. In this way, company A can search through its node and verify that B has a corresponding DUNS number and is a bona fide company. It is to be noted this feature is only available for D&B blockchain consortium customers.

The value of such a blockchain can be extended by connecting the ERP system of company A with the D&B blockchain consortium. This enables automatic verification whenever a new customer or supplier needs to be on-boarded. The D&B blockchain consortium is built on the open source Ethereum blockchain platform. This allows trading partners to confirm their identities and the terms of contracts through secure and immutable digital records that are stored on D&B's blockchain. The blockchain system matches the inquiry of the partner using DUNS numbers to a unique blockchain identifier code that is created for each company. It is to be noted that existing feature of D&B authentication through the API will still be available. The intent of moving to blockchain based authentication is to reduce the cost of authentication. Connecting it to ERP may provide further cost advantage.

Clearly, blockchain technology can help companies standardize data, connect ERPs, and create transparent consortiums with better security and authentication procedures. Blockchain will revolutionize how companies operate and maintain master data of items, supplier names, and customer names. In blockchain, every node is up-to-date, and the data is immutable. This drives data standardization, transparency, synchronization, and security across business entities and eliminates the risk of data duplication.

4. BLOCKCHAIN-DRIVEN ENGINEERING DESIGN

Engineering design is about using tools, techniques, and scientific knowledge to solve technical problems. Today, there are specialized firms that provide engineering design solutions in their areas of expertise. While these institutions store high volumes of propriety information, data is often tightly controlled and communicated only through people or technology interfaces. This process involves a lot of bureaucracy. To combat this, the engineering design sector has developed design software that integrates with ERP systems to transfer data for business transactions.

A typical example is the CAD/CAM system that integrates with the ERP system and transfers bills of material and routing data to ERP. Other examples include quality monitoring system, 3D printing, technical product evaluation system, etc., that also integrate with ERP systems. As these systems operate independently, integrating them with blockchain as well as ERP systems will provide significant advantages. One example of such integration is SAP Leonardo that provides customers and partners with access to standard SAP products that are enhanced and augmented with blockchain functionalities.

Here is how blockchain can be used for engineering design:

- Single platform: In engineering design, blockchain can unify customers and partners across the globe onto a single platform, thereby cutting across ERPs. This simplifies identity ownership, verification, authenticity, data provenance, sourcing, and reconciliation processes. Besides accelerating design and prototyping processes, it will also encourage joint manufacturing.

- Certifying agents: In the manufacturing sector, there are third party agencies, or certifying agents, that provide verification services, benchmarking services, certificates, and material identification services. One example is the National Center for Manufacturing Sciences that matches companies with complimentary abilities. These certifying authorities/agencies share data as physical or even digital copies, which makes auditing cumbersome and time-consuming. In cases where quality issues arise, the timely availability of these certificates and their authenticity become extremely critical. When ERP systems of the certifying agencies are connected through blockchain, it greatly simplifies such transaction processes by enabling transparency.

French automaker Renault [9] is already using a Blockchain solution based on Microsoft Azure to manage car ownership and maintenance details, track its products in the market, and simplify the release of new designs. Typically, maintenance records of an automobile usually reside with various repair shops, insurers, or dealers. This makes it difficult for original equipment manufacturer (OEMs) to push new changes or designs to existing models or sold cars. In case of product recall, the manufacturer is forced to depend on dealers as they have no direct access to customer information. Blockchain not only solves these issues but allows manufacturers to push newer designs and track products in the market.

5. BLOCKCHAIN-DRIVEN ORDERING AND PROCUREMENT

5.1 Procurement

Procure-to-pay or source-to-settle is a core procurement process involving sourcing decisions, supplier selection, supplier on-boarding, purchasing, receiving, supplier invoices, and payments. This process extends beyond the procurement division to include the finance, warehousing, and legal departments. Typically, all these processes occur in an ERP system involving data transfer and interaction across trading partners.

Today, blockchain can significantly change how these transactions take place. For instance, legal and procurement teams negotiate with trading partners to establish standards, terms, and conditions, transferring data through the system. Such data is stored either electronically or on paper. By enabling smart contracts between trading partners, blockchain can provide a digital and automated record that is constantly updated with the status of sold and delivered goods. It can establish rules and enable authentication, all of which will be stored on the blockchain. Thus, blockchain can also serve as a single repository for data exchange and invoicing.

Here is how blockchain can address common procurement challenges:
- People, places, and processes: These are the three major gaps in procurement. When a product arrives at the customer's location, the customer has no clear information about where all the product has been through, who has handled it and what kind of processes were involved in manufacturing it. Blockchain can store product information across the procurement lifecycle such as how it was made, who made it, where it was made, and product provenance. This provides customers with track-and-trace capabilities, enabling better visibility, greater control, lower risk, and improved regulatory compliance. This is detailed out further in the product provenance section.
- Standardizing terms: Most payment terms, freight, and discounting terms are not standardized across trading partners. This means that organizations must juggle several terms for different partners in various geographies, categories, and business lines. Blockchain can streamline this task by maintaining smart contracts. This will eliminate the need for multiple

data maintenance and there will be one authentic centralized data repository of terms and conditions accessible to all partners.

- Seamless integration: Disputes on supplier invoices for delivered goods are very common and key concern across enterprises. Gartner and J.P. Morgan estimate that, on an average, 10%–40% of all supplier invoices are disputed [10]. IBM Global Finances [11] survey shows that over US $100 million in invoices are in dispute between the buyer and supplier at any moment, requiring an average of 44 days for resolution. Blockchain can eliminate such causes for dispute by nearly 90%–95% [12–14]. Smart contracts allow details like proof of origin and delivery to be hosted in the immutable blockchain ledger, thereby authenticating demand, transfers, and transactions and reducing disputes. Also in case of disputes the data from blockchain (which is immutable) acts as the ultimate truth of evidence to easily settle the differences.

- Messaging system between partners: Typically, trading partners interact through email, EDI messaging, or B2B messaging systems. Blockchain can replace all these three systems by providing a unified business messaging system. While Margo [15] believes that blockchain is the new type of B2B standard messaging system, its success depends on widespread adoption. Through secure, auditable, and transparent information, blockchain can overcome the limitations of one-way, point-to-point, and batch approaches of existing messaging systems.

- Supplier benchmarking: Blockchain can help form consortiums in business-specific scenarios and share among themselves information like supplier performance, credibility, and quality aspects. This allows organizations to benchmark suppliers in the consortiums, enabling blockchain partners to make informed decisions about potential suppliers and future sourcing decisions.

Thus, blockchain can function as a supplementary layer between trading partners and their ERPs. Users can continue to use the ERP system even as the data is continuously transmitted to the blockchain digital ledger. The blockchain layer will improve trust and agility in the transaction and minimize disputes. Consortium-based supplier benchmarking is thus possible only through blockchain.

5.2 Ordering

Customer purchasing behavior is undergoing tremendous change thanks to internet and mobility. Customers, whether individuals or businesses, prefer to buy products online, on their mobiles or in-store as is accessible at the time of need. The challenge for an enterprise today is to make the product available to customer at his point of need in his desired location irrespective of how it was ordered. This is called omnichannel distribution. The customer experience has to be same irrespective of how it was ordered and how it was received like through shipments, in-store pickup, lockers, etc. For most organizations, managing and tracking orders is a full-time job for sales representatives. Without real-time inventory visibility, it becomes challenging to ensure accurate and timely order transmissions, acknowledgements, etc.

While a blockchain-driven ordering system may appear too futuristic, it can transform order management. The critical elements of an order, particularly cost, are driven by trust and assumed in real time, which leads to disputes. Similarly, order quantity is often sensitive. Blockchain can address these challenges by changing the mechanism of interaction. For instance, ordering processes can use private (permissioned) blockchains. Here, a new customer is evaluated by assessing their history, transactional and payment performance, DUNS verification, etc. He must be on-boarded onto the blockchain network before he can place orders.

In a permissioned blockchain, once a customer is on-boarded, he gains full visibility and real-time updates for price, product catalog, price breaks, etc., as this information comes directly from the blockchain network. Once the order is booked and confirmed by the customer, it is copied to all the nodes. Sensitive transaction details will not be available for other customers to see and can be veiled under metadata. This eliminates the need for EDI transactions and customer contacts, minimizing order transmission delays. Smart contracts between the customer and blockchain network initiator or even other members will drive the mechanism for ordering, order change, order promising, fulfillment logistics, and other aspects of order management.

6. BLOCKCHAIN-DRIVEN DEMAND AND SUPPLY MANAGEMENT

6.1 Demand Management

With the advent of collaborative planning, forecasting, and replenishment, the supply chains echelons today are integrated and interdependent. In an integrated supply chain, the forecast is shared across partners to improve supply planning. Most standard ERP products provide supplier scheduling/ scheduling agreements that are widely used across organizations to share demand across partners. Such sharing is enabled through methods such as:

1. EDI transactions: Customers either use the planning schedule transaction, (EDI code 830) or the shipping schedule transaction (EDI code 862). Planning schedule transactions are like forecasts, while shipping schedules are the same as confirmed orders.

2. Excel sharing: Many companies use a predetermined Excel format to share demands among partners. These Excel files are generally uploaded into the transaction system to improve supply chain or MRP.

3. Direct maintenance in the source system as per customer forecast (ERP feature): Many ERP systems provide customers with a specific login to access the ERP system to upload, maintain, change, and confirm demands. These demands are subsequently used to drive supply chain or MRP.

As sharing data is limited to these three options, there is no real-time integration. Data must be exchanged several times before it is finally approved for use in supply planning. This creates confusion regarding the final data to be used, resulting in the need for additional processes like forecast approval. Further, this approach does not offer demand visibility to supply chain partners within a company and across companies. Companies that want to adopt lean SCM philosophies must synchronize data, production schedules, and demand requirements among all partners so that they can respond to demand changes and surges. However, this requires real-time data visibility, trust, rules of engagement, and above all, the right technology to support these interactions.

In this scenario, blockchain can deliver true transformation by improving how these transactions are executed. Here, open blockchain is unsuitable as data is shared among a limited number of trusted partners. So, a permissioned blockchain is a better choice. Partners and their relationships must be verified before they are on-boarded to the permissioned blockchain. Smart contracts can be used to set up the rules of engagement and share trusted data in real time. Such real-time data sharing across partners enables a robust consensus-driven causal forecast apart from statistical forecasts, thereby improving forecast accuracy. This eliminates the need for forecast approvals and reconciliations, enabling organizations to respond intuitively and faster to dynamic supply chain demands.

6.2 Supply Management

Today, the process of supply involves multitiered suppliers, manufacturing sites, contract manufacturers, distributor warehouses, regional, national and local distribution centers, and scattered inventory across third-party logistics providers (3PLs). The cost of inventory is a concern for all industries [16]. With the advent of omnichannel distribution, the marketing, sales, and distribution of products become extremely complex and involve higher number of business partners and associates. This results in poor supply visibility.

To adopt lean philosophies such as zero inventory, just-in-time (JIT) manufacturing, and single piece flow, inventory must always be available in the required quantity. This mandates deep trust among partners and the assurance of product availability at the right time and in the correct quantity. Failing to meet these demands can result in tough penalties. By ensuring supply commitments and supply visibility across manufacturing sites and contract manufacturers, companies can drive supply chain transparency, thereby reducing administrative effort, cost, and counterfeit products.

Open and permissioned blockchains can enable these high level of supply visibility. As there are limited stakeholders to share the supply data, a permissioned blockchain is more suitable. In a permissioned blockchain, all stakeholders can be part of the node that shares information when needed. Such data sharing can be set up and managed through smart contracts where the required node shares relevant information in adequate frequency. The best way of sharing information is by connecting the transactional ERP system to the blockchain. Adopting blockchain for supply visibility through the connected ERP system delivers significant benefits such as:

(a) Reduced counterfeiting: Providing product provenance details helps partners trace the origin of the product, its ingredients, ownership, and storage details, thereby eliminating counterfeit products.

(b) Enabling digital: Product details and its lifecycle are stored within the system in a digital format, eliminating product ambiguity.

(c) Improved sourcing: While implementing blockchain among all partners will be challenging, particularly for multiple and multilayer suppliers, it will drive long-term benefits of transparency, sustained growth, and accountable and responsible sourcing.

(d) Faster operations: Transparent auditing improves compliance with governmental regulations and accelerates customs clearance. Blockchain will eliminate the need to file for country of origin reports and other customs/border clearance papers. All these information will be readily available in blockchain for instant access by governmental agencies. An example here is the Port of Rotterdam Authority in Netherlands that has launched a field lab to explore blockchain technology. The idea is to use blockchain for customs scanning to reduce vessel turnaround time in the port.

(e) Higher growth: Transparency among all stakeholders and partners will drive sustained growth with immutable, secure, and mutual data exchange.

(f) Faster supply availability: As higher number of enterprise and supply generating systems like ERP, MES, etc., get connected to blockchain, data availability becomes more real time and transparent.

Thus, blockchain can provide macro- as well as microvisibility across multiple suppliers from agri produce goods to high-tech manufacturing. It can also integrate with ERP to provide seamless data exchange and stakeholder interactions.

7. BLOCKCHAIN-DRIVEN MANUFACTURING AND LOGISTICS MANAGEMENT

7.1 Manufacturing Management

The Blockchain Research Institute states that blockchain is an Industry 4.0 technology [17]. In the news article, Neil [17] explains that Foxconn is using blockchain to simplify integration with its partners and streamline the movement of money and goods. With the proliferation of contract manufacturing, particularly in high-tech manufacturing, blockchain will play a crucial role by reducing the dependency on EDI transactions and enabling real-time updates on demand, supply, manufacturing status, and availability information.

To explore the application of blockchain in manufacturing, let us first understand the key pain points of the manufacturing industry today. Duvall [18], Carr [19], and Handfield and Steininger [20] list that the key challenges faced by manufacturing organizations with an existing ERP system are final product quality, counterfeit components, meeting governmental regulations, staying competitive with better products at lower cost, ensuring supply chain transparency, controlling cost, dealing with fragmented supply chains, managing multifacility and duplicate data/records, enabling data visualization and intelligence across entities, and driving collaboration among suppliers or contract manufacturers. A closer look at these challenges shows that the inherent capabilities of blockchain can be used to deliver significant benefits. The main question is: how can blockchain address these challenges for manufacturers? The subsequent section provides an answer to this pertinent problem.

In manufacturing, a permissioned blockchain is more suitable than an open blockchain. A permissioned blockchain will be more acceptable and beneficial to the manufacturer as it gives control to on-board only specific supply chain partners as a node. The blockchain can act as a centralized distributed ledger that is owned and initiated by a manufacturer. It will have the capability to seamlessly collate data from multiple ERPs belonging to customers, suppliers, distributors, and contract manufacturers. This data can then be used by the original manufacturer for decision-making, transparency, and status updates. For example, an OEM may want to on-board only its top 10 suppliers and all contract manufacturers onto the blockchain. This will give him better control and in-depth visibility into material movement and manufacturing, which comprises the bulk of his business. In this way, blockchain addresses the challenge of fragmented supply chains.

In order to be audit-compliant, manufacturers need certificates that are usually obtained after a lot of paperwork, emails, phone calls, site visits, surprise checks, followups, etc. This gives rise to third-party certifying agencies like the International Organization for Standardization and BIS Hallmark, a jewelry certifying agency in India. However, manufacturers often lack visibility into the status and operations of a contract manufacturer and are mostly blind to its sourcing. Blockchain can help manufacturers gain better control over governmental regulations as all legal transactions are cryptographically stored in the blockchain. Any auditor or government body can access the blockchain node on the distributed database and use it for the intended purpose. Immutable data stored on the blockchain simplifies auditing, thereby reducing cost. Blockchain can also leverage smart contract rules to reduce nonapproved suppliers, products (ingredients), country of origin and transactions in ERP.

Additionally, blockchain can help manufacturers innovate their product lines faster and at lower cost, control counterfeit components, and combat supply chain transparency issues. This is possible because blockchain supports secure sharing of design documents across partners (suppliers, contract manufacturers, and OEMs), thereby enabling visibility into status of prototyping and sourcing and allowing frequent review of sourced costs. Blockchain can also track the cost of products from multiple sources to control cost of the end-product. This will promote trust and secured sharing leading to accelerated innovation unconstrained by geographical boundaries. It will also drive economic development in emerging markets. Faster product development will improve the speed of material flow and reduce localized inventories, thus creating newer markets and opportunities.

7.2 Logistics Management

The logistics industry is currently undergoing a transformative change. Consider how Uber now delivers food, electric cars are replacing fuel-based vehicles, and flying cars are almost a reality. Technology is also transforming the logistics industry by enabling drone delivery. These futuristic solutions primarily benefit end customers and individuals. However, corporate logistics functions still struggle with transparency issues, high cost, manual effort, lack of trust, and challenged by real-time data availability. Most logistics executives focus on dealing with crises, disruption, and technology rather than collaborating with the freightliners.

According to research and articles published by Kavas [21], World Economic Forum [22], Morales [23], and Brosch [24], the logistics industry is plagued by challenges of risk, variability, volatility, and disruption. The reasons for these are natural disasters, violence, noncompliance with government regulations, poor accountability, delayed information sharing for logistics, poor communication among partners, limited shared logistics capability and capacity utilization opportunities among service providers, managing inventory visibility across various logistics providers, theft and pilferages, and the need for segmented and customized services as demanded by customers. As per Kralingen [25] the information flow in logistics industry is highly inefficient, error-prone, manual, nondigital, and heavily dependent on complex paper-based systems.

Today, the logistics industry needs better technology, efficient route planning, and stronger integration with ERP systems. While blockchain cannot provide these capabilities, it can ensure more transparency, reduce or remove dependencies on intermediaries, enable inventory visibility across the supply chain, and drive effective tracking and communication. These capabilities can significantly improve the functioning of the logistics industry.

To improve transparency and remove dependencies on intermediaries, Maersk and IBM have jointly developed a global trade digitization platform based on blockchain. Port Houston, Rotterdam Port Community System, the Customs Administration of the Netherlands, and the US Customs and Border Protection will be part of this blockchain network [26]. Any company using this blockchain will enjoy paperless and seamless transfer of goods from one location to another. This is possible because the customs organizations are a node on this network and can identify what, when, where, and how much is being transferred. Similarly, in 2017, Samsung launched a blockchain consortium comprising the Korea Customs Service and Korea's Ministry of Oceans and Fisheries. Another example is the Port of Antwerp in Belgium that is testing the use of blockchain to automate and streamline container logistics operations in its terminal.

Thus, in the last few years, there has been a rise in logistics-specific blockchain solutions. For instance, the Blockchain in Transport Alliance (BiTA) [27] provides a platform to standardize and develop transport-related blockchain solutions. Some of its members include UPS, Fedex, SAP, BNSF, Salesforce, Schneider, JD.com, and Penske. Fedex was one of the first companies to join the forum and help develop blockchain technology standards and education for the freight industry. As more logistics giants join the blockchain platform, it will enrich the platform and enable smaller players to use these platforms for higher benefits.

The logistics industry can derive greatest value from blockchain when it connects logistics partners and manufacturers' ERP systems to the blockchain network. This will enable seamless information flow from one system to the centralized distributed ledgers, which can then be shared with any of the partners who are a node in the network. These blockchain-connected ERP systems of logistics service providers can improve supply chain functioning across the whole industry, bringing in transparency for customers and simplified product provenance tracking. Further, the digitization of logistics-connected ERP will replace paper-based information transfer through digital capabilities and help track and reduce pilferages and theft.

Blockchain will smoothen and accelerate the process of customs clearance, vessel turnaround time, and auditing. It will also improve visibility across supply chain partners, enabling near real-time tracking, and updates in case of supply chain disruption. When natural disasters occur, blockchain can check for alternate routes, suppliers, products within the network to make inventory/material available. Further, it provides a platform to share container capacity, port capacity, route capacity, warehouse capacity, and other assets. Finally, coupling blockchain with digital transaction will reduce paper usage, driving green initiatives to protect the planet.

8. PRODUCT PROVENANCE USING BLOCKCHAIN

In 2015, diners at a US-based fast food restaurant contracted an *E. coli* infection. Despite independent reviews by state and federal regulatory officers as well as detailed independent investigation by various investigating agencies the company filed to stock exchange that there was no single food or ingredient that caused the infection. It is speculated that products like beef and agri produce like cilantro caused the outbreak, but sources are still unknown. In another example, Apple came under pressure from various organizations to audit its entire supply chain and confirm that chemicals used in manufacturing such as tantalum, cobalt, gold, tin, and tungsten were conflict-free and not sourced from companies that funded armed groups or violated human rights [28]. Steve jobs admitted this was challenge stating, "Until someone invents a way to chemically trace minerals from the source mine, it is a very difficult problem" [28]. These examples highlight the pressing need to know the origins of a product and trace its lifecycle from source to consumer, a process known as product provenance. The importance of this is significant when one considers how Mattel, the maker of Barbie dolls and Hot Wheels cars, had to recall around 1 million toys that had lead paint. From tracing non-GMO ingredients in the food supply chain to knowing whether leading apparel brands source from manufacturers who abuse human rights, product provenance is critical to supply chains and extremely challenging.

Companies must trace their ingredients or components to protect their reputation, ensure component quality, and inform the customers about the quality and authenticity of their finished products. However, supply chain or ERP software cannot provide such information as ERP is confined to a specific company or enterprise. The modern supply chain has become increasingly complex involving multiple parties/stakeholders across the world. Thus, companies need a centralized software platform that leverages a decentralized approach to connect disparate systems across the globe, irrespective of location or technology used.

Blockchain offers this very functionality. Product provenance can be used for a variety of products like packaged food, produce goods, seafood, diamond jewelry, precious metals, garments, rare earth elements, and fashion products to name a few. The use of blockchain to enable product provenance is seeing rapid adoption through two main types of solutions:

(1) Independent tracking via blockchain: Companies such as Provenance, Sourcemap, and Owlchain provide independent product tracking right from raw materials/origin to the end product that is delivered to the consumer. Provenance works on Bitcoin as well as Ethereum platforms and is a complete blockchain solution. Sourcemap is an interactive mapping platform that uses a global map to show users where the various elements came from. Now, Provenance and Sourcemap have linked their digital platforms to benefit customers, which is a breakthrough.

(2) Custom-built provenance solutions: Software service providers can use the blockchain framework to build provenance solutions for its customers (permissioned blockchain). For instance, Infosys has developed a product provenance solution using Oracle Blockchain Cloud services that is based on Hyperledger fabric [12–14]. Infosys has also developed a coffee bean tracking provenance solution for its customers. These examples prove that there is a need for custom built provenance solution which can be developed with product or industry specific validations.

It is important to note that the idea of provenance can only work when all the supply chain stakeholders are a part of the blockchain network. The architecture of blockchain inherently traces products as they pass from one supply chain entity to another. These transactions are stored as blocks and are chronologically linked according to the physical movement of the goods.

A fitting example of supply chain provenance and visibility is the ethical seafood traceability solution provided by Hyperledger. The Hyperledger Sawtooth framework records the journey of seafood from its origin, i.e., where it is caught till the end consumer. Consumers can view all this information on their smartphone using a mobile application. When seafood is caught, it is attached with IoT sensors that track its movement during transportation. The tracking mechanism monitors ownership, possession, location, temperature, humidity, motion, and shock among other things. Another example of product provenance is the Everledger Blockchain that provides provenance for diamonds. With more than 1.6 million diamonds on its blockchain network, it creates a digital record of a physical diamond by capturing attributes like color, carat, and certificate number, which is etched on the diamond through lasers [29]. The goal of such solutions is to provide consumers with instant information about the source of products and their physical movement.

With growing customer demand to know the source and path of products, it will be imperative for manufacturers to enable supply chain transparency and product provenance. Several multinational companies and large retailers are already planning to adopt provenance solutions. Soon, this will become the new norm in modern supply chains, and blockchain is a promising and acceptable technology to achieve this. The connected ERP's to the centralized blockchain system will definitely ease the information flow and provenance.

9. BLOCKCHAIN USE CASES IN THE DISTRIBUTION INDUSTRY

So far, this chapter has dealt mostly with how blockchain can impact the operations of contract and OEMs. Another key area that deserves mention is the distribution industry. This industry is witnessing tremendous transformation because of newer technologies, and blockchain has a significant role to play. Blockchain is particularly important for this industry as its very dynamic with multiple stakeholders in the form of customer, supplier, manufacturer and 3rd party service providers. Blockchain provides capabilities that are set to improve operations, communications, transparency, visibility, and the bottom line of the distribution industry.

Before delving into the benefits of blockchain, let us evaluate the challenges in the global distribution industry. Research and analyses by Mulky [30], Yu et al. [31], Arnold [32], and Bolduc [33] show that the main challenges of distribution industry are:

- Synchronizing customer demand and supplier inventory: Distributors often struggle to effectively service the needs of customers and communicate with them seamlessly and transparently. To serve customers, the distributor needs to source, communicate, and receive materials from the suppliers. Here, the challenge for a distributor is in meeting customer demand for faster service and cheaper products, and sourcing it in this way from the supplier. Customers want to see the best price with discounts from the right source at the very beginning. They want information on shipment, transit time, lead time, acknowledgment, and a proforma invoice as soon as they place an order along with the option of changing their order anytime. Similarly, distributors expect to receive inventory from their suppliers on time with the option to change their order anytime. Thus, customers, suppliers, and distributors want real-time information on order status, transit visibility, and pricing visibility.

- Transparency of sale from supplier, distributor to end customer: Distributors are legally bound to provide information of sale to their suppliers source of material information to their customers. This information is also needed to enable, manage, and check the ship and debit claims. Such information enables visibility and profitability analysis of distributor channels and price management in the market. In a multilevel distribution channel, this information exchange can become extremely complex with limited visibility. While ERP systems provide some visibility, it is not in real time and is not across all stakeholders. The main reason for this is that ERP systems are not connected with each other and confined to an enterprise. There are a number of reports shared today across suppliers, manufacturer and distributor like Point of Sale (POS) report, Product Transfer and Resale Report, Availability report, Planning and Shipping schedules, etc. which are time consuming and asynchronized.
- Counterfeit products: Counterfeit products invariably create revenue loss for distributors because of higher instances of product returns. Further, in some cases, customers transfer the financial liability caused by faulty products to its distributors. Without proper insurance, this can have significant negative impact on a distributor's reputation and credibility in sourcing the right product.
- Item information and its attributes (for informed buying/sourcing): Google has created a breed of tech-savvy customers who can instantly access information on-the-go. Such customers want exact information on the product being sold. According to Banerjee [12–14], companies are expending significant effort in providing customers with granular information from nutrition facts to carbon labels. However, this requires product provenance and detailed product information.
- Moving to an omnichannel distribution supply chain model: Today, most retailers and distributors are shifting to an omnichannel distribution supply chain model. The intent of the distributor is to use existing stock to fulfill orders from retailers and customers across various channels like stores, franchises, online, and mobile. This helps distributors control price and stock, improve margins, increase their customer base, and boost revenue. However, it also increases supply chain complexity.

The challenge here is to ensure minimum turnaround time while seamlessly connecting customer demands and supplier purchase orders irrespective of location.

Blockchain can help solve most of the above problems. The blockchain can be centrally owned and managed by distributors. Using several methods, suppliers and customers can be a node in the blockchain network, thereby promoting transparency. Alternatively, the distributor can be a node in the manufacturer/supplier or customer-owned and managed blockchain network. In both these cases, the blockchain can be either open or permissioned networks.

We have already discussed several capabilities that are possible in blockchain. Based on these, product provenance and supply/demand visibility will resolve the challenge of counterfeit products, item attribute information, and synchronizing customer demand/supplier supply. A blockchain that is synchronized and integrated with the ERP systems of distributors, suppliers, and customers can provide visibility into order and inventory status and help with ship and debit claim management. ERPs connected to the centralized blockchain network will deliver the speed and flexibility necessary to support omnichannel distribution.

Here, a private or permissioned blockchain is optimal since the network only contains distributors and its suppliers and customers. Permissioned networks do not require miners. The blockchain network will contain the complete ledger information of inventory, orders, in-process inventory, open orders, backlogs, and forecasts. Further, each customer order can leverage smart contracts to understand customer preferences such as branch, area, location, or supplier preference based on type of order. Smart contracts will also help identify the right discounts for the customer. Finally, integrating customers, suppliers, and distributors will provide real-time visibility of price and availability.

The challenges are different for specialty distributors such as those who handle electrical equipment, commodity products like metals, wired products, autocomponents, and home essential products. These distributors face challenges of identifying the correct part (item), the right product manufacturer/supplier, the right price (pricing agreements) of an item, product origin, and calculating the right claim (ship and debit claim). Here, the root cause is asynchronous data between the distributor, manufacturer, and customer and the lack of a central governing system.

While electrical distributors use independent third parties like Trade Service (a Trimble Company) or Industry Data Warehouse (IDW) to gain data for item (and its related information ranging from physical characteristics to technical information), cost, and price details in electronic formats, such agencies do not offer information on product origin. In cases where data is available, the challenge lies in regularly updating these data feeds into the transactional system of distributors (or any consumer of the data). This underscores the need for a centralized system for real-time data.

Blockchain can address these challenges by synchronizing data among the manufacturers, distributors, and suppliers. This will accelerate order processing, enable accurate pricing and discounts, identify product provenance, and, most importantly, identify the right item through technical attributes. Blockchain is particularly useful for warranties as there is no need to store the installed base data in individual ERP systems. Instead, the immutable transactions in the blockchain network will provide details of the purchase and available warranty for the product, and this information will be accessible to manufacturers, distributors, and customers.

Microsoft has partnered with Mojix to leverage Azure and provide blockchain-as-a-service solutions for retailers and distributors [34]. Mojix is a leading provider of RFID and IoT solutions. Through this partnership, suppliers/distributors get access to location-based and real-time data of the status of inventory even when it is in transit. Materials are scanned when they pass certain check points and these insights are captured using Mojix technologies and pushed to the Azure blockchain, thereby providing immutable and real-time data. Such visibility helps distributors track orders and get real-time updates.

10. CHALLENGES AND FUTURE OUTLOOK OF SUPPLY CHAIN WITH BLOCKCHAIN AND ERP

ERP has successfully integrated departments and operating entities within an enterprises. The adoption of blockchain will, over time, provide the opportunity to connect individual ERP systems to the blockchain. The future of the supply chain is one where blockchain will be able to track activity beyond an enterprise's boundaries. While such capabilities will transform supply chain functioning, it comes with its own share of challenges. It is important to remember that the present challenges in adopting a technology may evolve into a different set of challenges as the technology itself evolves. Currently, the key challenges in adopting blockchain for supply chains are:

1. Infrastructure and network: Blockchain needs online transactions, communication, and data storage. All of this consumes significant internet

bandwidth and CPU power, which can be challenging in developing and underdeveloped economies.

2. Interoperability: As highlighted by Banerjee [12–14], blockchains should seamlessly work across ERP and other blockchain systems. Only when blockchains are allowed to communicate with each other and transfer data/information to other ERP/transactional systems can companies or supply chain realize the tangible benefits and full potential of the technology.

3. Costs of on-boarding and maintenance: In a blockchain-enabled supply chain, the nodes are supply chain partners, i.e., suppliers, customers, distributors, and manufacturers. On-boarding and maintaining these nodes, particularly for high volume transactions, are a significant cost and must be laid out clearly in smart contracts.

4. Data storage cost on blockchain (data per transaction): Blockchain data is stored on the cloud. A blockchain database must store data indefinitely and such indefinite storage of data involves high cost. This will require a different type of storage model as recurring payment models will not work. The indefinite retention of data is costly and data models has to evolve for these. These are going to cost and impact every node members.

5. Data validation latency: In an open network blockchain, the blocks of data are queued and validated by miners for every transaction. There is a latency before the block is verified and becomes available as a valid block in every node. In a Bitcoin network, the average latency ranges from 8 to 19 min, but this can even increase to hours. From a transaction perspective, considering these are real life transactions with business decisions based on it these values are quite high. Ideally, the latency must be in seconds if blockchain is to be a viable solution for supply chain transactions with corresponding business decisions and actions based on it.

6. Payload size restriction: There is a payload size restriction for any blockchain network whether open or private. As the payload is not infinite, there will be transactions that become invalid due to size. While the limit for Bitcoin is roughly 1 MB, there is no standard for payload size as it depends on the network's capability of nodes. This raises concerns about network and infrastructure challenges when it comes to payload restrictions. While this challenge may change over time, currently it is a key concern.

7. Regulatory and legal acceptance: Blockchain has no legal framework. Countries vary on their view of Blockchain and cryptocurrencies; while some are excited, others are suspicious. In the United States, cryptocurrency is regulated by state jurisdiction. The EU is open to blockchain and cryptocurrency. However, the lack of a global regulatory framework poses challenges for global acceptance and adoption.

8. Trust: This is the last and the biggest concern from a supply chain perspective. Blockchain can improve visibility, real-time tracking, and seamless transfer of data. However, the central design element for blockchain is implicit trust between partners. For blockchain to be a part of supply chain transactions, every node (suppliers, customers, distributors, and manufacturers) should be open to sharing their data.

Most of these challenges revolve around cost, network infrastructure and legal aspects. These may change over time as technology makes progress and laws change and adopt blockchain technology. But those will bring forward a new set of challenges. Similarly, the challenges in supply chains themselves may transform in future. For instance, data security in blockchain-enabled supply chains is a concern as data is shared outside the enterprise. While blockchain is designed to be robust, secure, and anonymous, there have been incidents where it is misused by hackers who seek ransom payments in bitcoins. There have also been reports of hacking in cryptocurrency exchanges. Even though these incidents do not directly impact supply chain operations, we can expect that innovation in blockchain networks will address these loopholes and improve security.

11. CONCLUSION

Blockchain networks can be connected to ERP systems within and outside an organization to facilitate information and transaction sharing across enterprises. As a transparent and decentralized platform, blockchain can complement and simplify supply chain operations by enabling connected supply chains that share information across all transactions. Blockchain is set to revolutionize various aspects of daily life as it provides consumers with more information on the products they buy through product provenance. It will encourage the formation of consortiums where companies can improve their supply chain operations. Some of the benefits of blockchain-based supply chains include improved supply chain decision-making, sourcing, inventory tracking, transparency, and visibility. These benefits not only improve supply chain operations but reduce cost and time.

When coupled with the Internet of Things, blockchain can drive instant sharing of real-time data and accurate updates. In future, we can expect blockchain to bring people and countries closer, making this world a better place to live in.

REFERENCES

[1] H. Ince, S.H. Imamoglu, H. Keskin, A. Akgun, M.N. Efe, in: The impact of ERP systems and supply chain management practices on firm performance: case of Turkish companies, 9th International Strategic Management Conference, Procedia—Social and Behavioral Sciences, 99, 2013, pp. 1124–1133.

[2] W. Hwang, The Drivers of ERP Implementation and Its Impact on Organizational Capabilities and Performance and Customer Value, Ph.D. Dissertation, The University of Toledo, 2011. p 284.

[3] A. Banerjee, Information technology enabled process re-engineering for supply chain leagility, International Journal of Information Technology and Management 14 (1) (2015) 60–75.

[4] S. Li, B. Ragu-Nathan, T.S. Ragu-Nathan, S. Subba Rao, The impact of supply chain management practices on competitive advantage and organizational performance, Omega 34 (2) (2006) 107–124.

[5] C. Batini, M. Scannapieco, Data Quality: Concepts, Methodologies and Techniques (Data-Centric Systems and Applications), Springer-Verlag New York, Inc. Secaucus, NJ, USA, 2006, pp. 1–18.

[6] O. Foley, M. Helfert, Information quality and accessibility, in: T. Sobh (Ed.), Innovations and Advances in Computer Sciences and Engineering, Springer Netherlands, Dordrecht, 2010, pp. 477–481.

[7] T. Knap, M. Nečaský, M. Svoboda, A framework for storing and providing aggregated governmental linked open data, in: A. Kő, C. Leitner, H. Leitold, A. Prosser (Eds.), Advancing Democracy, Government and Governance. EGOVIS/EDEM 2012, Lecture Notes in Computer Science, vol. 7452, Springer, Berlin, Heidelberg, 2012.

[8] B. Levine, Dun & Bradstreet Is Testing Blockchain as a Way to Securely Distribute Its Content, Martechtoday.com. https://martechtoday.com/dun-bradstreet-testing-blockchain-way-securely-distribute-content-202459, 2017. Accessed 14 January 2018.

[9] S. Higgins, Automaker Renault Trials Blockchain in Bid to Secure Car Repair Data, https://www.coindesk.com/automaker-renault-trials-blockchain-bid-secure-car-repair-data/, 2017. Accessed 12 February 2018.

[10] Fidesic Corporation, The True Costs of Invoicing and Payment, http://www.enlivensoftware.com/Portals/0/docs/Cost%20of%20Invoicing.pdf, 2002. Accessed 12 January 2018.

[11] IBM Blockchain, IBM Global Financing Uses Blockchain Technology to Quickly Resolve Financial Disputes, https://www.ibm.com/blockchain/infographic/finance.html. Accessed 23 February 2018.

[12] A. Banerjee, Integrating Blockchain With ERP for a Transparent Supply Chain, https://www.infosys.com/Oracle/white-papers/Documents/integrating-blockchain-erp.pdf, 2017. Accessed 1 January 2018.

[13] A. Banerjee, Product Provenance and Supply Chain Transparency Using Oracle Blockchain Services—An Infosys Offering!, http://www.infosysblogs.com/oracle/2017/10/product_provenance_and_supply_.html, 2017. Accessed 5 February 2018.

[14] A. Banerjee, Re-Engineering the Carbon Supply Chain With Blockchain Technology, https://www.infosys.com/Oracle/white-papers/Documents/carbon-supply-chain-blockchain-technology.pdf, 2017. Accessed 10 February 2018.

[15] T. Margo, The Blockchain Impact: How It Will Change Your B2B Network, https://www.ibm.com/blogs/watson-customer-engagement/2017/06/19/blockchains-impact-b2b-networks/, 2017. Accessed 5 January 2018.

[16] J.M. Castro, Can Blockchain and Cognitive Analytics Yield Higher Demand Forecasting Accuracy?, IBM Electronics Industry Blog, 2016. https://www.ibm.com/blogs/insights-on-business/electronics/demand-forecasting-reinvented/. Accessed 15 January 2018.

[17] S. Neil, Blockchain Meets the Manufacturing Supply Chain, https://www.automationworld.com/blockchain-meets-manufacturing-supply-chain, 2017. Accessed 11 January 2018.

[18] S. Duvall, Solutions to 5 of Manufacturing's Biggest Pain Points, http://www.loganconsulting.com/Blog/articleType/ArticleView/articleId/938/Solutions-to-5-of-Manufacturings-Biggest-Pain-Points#.WnWvZLynHIU, 2016. Accessed 3 February 2018.

[19] J. Carr, Manufacturing ERP: What Are Your Pain Points in 2014? https://ultraconsultants.com/manufacturing-erp-pain-points-2014/, 2013. Accessed 23 January 2018.

[20] R.B. Handfield, W. Steininger, An assessment of manufacturing customer pain points: challenges for researchers, Supply Chain Forum. An Internation Journal 6 (2) (2005) 6–15.

[21] S. Kavas, The 5 Biggest Problems of Global Logistics, https://www.morethanshipping.com/the-5-biggest-problems-of-global-logistics/, 2015. Accessed 28 January 2018.

[22] World Economic Forum, World Economic Forum White Paper Digital Transformation of Industries: Logistics, http://reports.weforum.org/digital-transformation/wp-content/blogs.dir/94/mp/files/pages/files/wef-dti-logisticswhitepaper-final-january-2016.pdf, 2016. Accessed 30 January 2018.

[23] J.D. Morales, Logistics & Transportation Executives Facing Today's Challenges, Seek Solutions Well Into the Future, http://www.stantonchase.com/wp-content/uploads/2015/07/Logistics-transportation-executives-facing-today's-challenges-seek-solutions-well-into-the-future.pdf, 2015. Accessed 1 February 2018.

[24] A. Brosch, 6 Global Supply Chain Challenges to Ignore at Your Own Risk, http://www.inboundlogistics.com/cms/article/6-global-supply-chain-challenges-to-ignore-at-your-own-risk/, 2015. Accessed 2 February 2018.

[25] B. Kralingen, IBM, Maersk Joint Blockchain Venture to Enhance Global Trade, https://www.ibm.com/blogs/think/2018/01/maersk-blockchain/, 2018. Accessed 4 February 2018.

[26] M. White, Maersk and IBM to Form Joint Venture Applying Blockchain to Improve Global Trade and Digitise Supply Chains, https://www.maersk.com/press/press-release-archive/maersk-and-ibm-to-form-joint-venture, 2018. Accessed 3 February 2018.

[27] D. Patel, UPS Bets on Blockchain as the Future of the Trillion-Dollar Shipping Industry, https://techcrunch.com/2017/12/15/ups-bets-on-blockchain-as-the-future-of-the-trillion-dollar-shipping-industry/, 2017. Accessed 5 February 2018.

[28] L. Browning, Where Apple Gets the Tantalum for Your Iphone, http://www.newsweek.com/2015/02/13/where-apple-gets-tantalum-your-iphone-304351.html, 2015. Accessed 6 February 2018.

[29] J.J. Roberts, The Diamond Industry Is Obsessed With the Blockchain, http://fortune.com/2017/09/12/diamond-blockchain-everledger/, 2017. Accessed 8 February 2018.

[30] A.G. Mulky, Distribution challenges and workable solutions, IIMB Management Review 25 (3) (2013) 179–195.

[31] Y. Yu, X. Wang, R.Y. Zhong, G.Q. Huang, E-commerce logistics in supply chain management: practice perspective, Procedia CIRP 52 (2016) 179–185.

[32] D. Arnold, Seven Rules of International Distribution, Harvard Business Review, https://hbr.org/2000/11/seven-rules-of-international-distribution, 2000. Accessed 10 February 2018.

[33] S. Bolduc, Top 3 Challenges Facing Industrial Distributors, https://www.spscommerce.com/blog/top-3-challenges-facing-industrial-distributors/, 2016. Accessed 10 February 2018.

[34] J. Schwartz, Microsoft Pitches Blockchain to Help Troubled Retail Supply Chains, https://redmondmag.com/blogs/the-schwartz-report/2017/01/microsoft-pitches-blockchain-to-retailers.aspx, 2018. Accessed 15 February 2018.

[35] S.C.L. Koh, A. Gunasekaran, D. Rajkumar, The involvement, benefits and impediments of collaborative information sharing, Int. J. Prod. Econ. 113 (1) (2008) 245–268.

[36] S.C.L. Koh, A. Gunasekaran, T. Goodman, Drivers, barriers and critical success factors for ERPII implementation in supply chains: a critical analysis, J. Strat. Inf. Syst. 20 (4) (2011) 385–402.

[37] R. Addo-Tenkorang, P. Helo, Enterprise resource planning (ERP): a review literature report, in: Proceedings of the World Congress on Engineering and Computer Science 2011 Vol II, San Francisco, USA.

FURTHER READING

[38] D.R. Robles, Blockchain Technology: Implications and Opportunities for Professional Engineers, National Society of Professional Engineers, 2016. https://www.nspe.org/sites/default/files/resources/pdfs/NSPE-Whitepaper-Blockchain-Technology-2016-final.pdf. Accessed 10 January 2018.

ABOUT THE AUTHOR

Arnab Banerjee is a Principal Consultant with Enterprise Applications Services of Infosys Ltd. He consults in the area of ERP and Blockchain for Supply Chain Domain. His consulting experience of more than 16 years spans across North America, Europe and Asia. In his blockchain consulting role he helps develop blockchain use cases and implement solution on blockchain products. His research interests include information technology and blockchain applications in supply chain management. He has extensive publications in the area of reverse supply chain, lean/agile/leagile initiatives, theory of constraints, supply chain transformations and humanitarian logistics. Over the years he has published more than 12 research papers in various peer reviewed international journals. He has to his credit eight international conference research papers, three case studies, and two book chapters. He has published Point of View documents, White Papers and blogs on Oracle products based on his experience and researches which he does while doing his consulting assignments. His blockchain publications are well read and widely regarded across the board. He holds a PhD in Supply Chain Management, a Master's in Industrial Engineering and a graduate degree in Mechanical Engineering. He is a certified Six Sigma black belt champion. He is listed in Who's Who of the World 2015 edition.

CHAPTER FOUR

Applications of Blockchain in the Financial Sector and a Peer-to-Peer Global Barter Web

Kazuki Ikeda*, Md-Nafiz Hamid[†]
*Department of Physics, Osaka University, Osaka, Japan
[†]Department of Veterinary Microbiology and Preventive Medicine, Iowa State University, Ames, IA, United States

Contents

Abstract

Recent advances in Blockchain technology has opened up a vast array of decentralized and distributed systems. The most significant contribution of this technology is it gets rid of the third-party functioning as a mediator in systems that requires trust for any kind of transaction. From this perspective, transaction of money is fundamentally an authorized third-party mediating trade of goods or services. In this chapter, we first review the conventional applications of Blockchain and next propose a novel economic system in which such value-added items are exchanged without using money and without going through a third party. In the long history of economics from ancient times, the monetized economy system today has a very short record, and it is essentially based on the

Advances in Computers, Volume 111
ISSN 0065-2458
https://doi.org/10.1016/bs.adcom.2018.03.008

existence of trustworthy central banks, governments, and organizations. However in line with the recent growing sense of decentralized systems and in keeping with unstable international affairs including the recent money crisis, it is meaningful to consider alternative economic system which can exist without going through any third party. To address this issue is one of the purposes of this chapter.

1. INTRODUCTION

A Blockchain is a decentralized system that uses peer-to-peer network to enforce transactional integrity as well as security, and gets rid of a third party such as banks for mediation. It offers a wide variety of uses, and the number of applications which use it to automate integrity of transactions for both monetary and nonmonetary purposes has increased over the past decade. As such it has attracted enormous international interest. Some countries are considering introducing the Blockchain technology to public services. Estonia has issued e-Residency, a digital ID that offers people all over the world to start a location-independent business. Their citizens can also use it for voting in elections, tax payment, bank transfer, and so forth. The potential of Blockchain for nonmonetary transactions and consequently shaking up every major sector is indeed immense.

We also propose a novel logistics system, Mariana, which is a decentralized global barter system. Barter is the starting point of all economics. In a barter system, transaction of goods happens directly for other goods rather than a medium of exchange such as money. Until the last century, it had been commonly and widely believed from ancient times that money was invented to solve inconvenience of barter. However, many economists and cultural anthropologists today are skeptical of this history between money and barter. What we aim at in this chapter is not to add another viewpoint toward this debate, but to give a solution to construct a real and global economic system which is truly based on barter. This can facilitate the exposure of a global nonmonetized economy that can thrive side by side of the traditional monetized economy.

Our system will be a peer-to-peer global barter system, which will make a part of the international economy get back to the basics, and maybe the first officially recorded economy which is comprised of a global barter trade. The core technology supporting Mariana is based on a Blockchain, a sequential distributed database in which transaction records are stored without going through a particular third party. In a sense, money is an authorized third party intermediating between goods or services, and a Blockchain removes such a third party from the existing system. This system will allow direct trades of products without going through a broker and without using money. In this system, importance of money will not be completely lost, and implementation of such a system will allow our society to have it both ways: monetized economy and nonmonetized economy.

2. APPLICATIONS TO MONETIZED ECONOMY
2.1 Peer-to-Peer Economy

Blockchain is a system where the need for a central authority to verify a transaction is eliminated. For example, when sending money to a friend, that transaction goes through a bank which mediates the transfer between the two persons, and updates the state of each person's bank account. Blockchain can eliminate the need for a bank as the transaction history is updated by a decentralized system which is open and transparent, and at the same time protecting the privacy of people. This immediately makes us think about the possibility of exchanging money and services without the need for a central entity, and the added cost that brings with it. Indeed, many companies are working on this. Circle [1] is a startup that provides a platform powered by Blockchain to receive and send money instantly across borders without any added fee. Billon [2] is another company that is trying to use the distributed nature of Blockchain to speed up transactions across countries.

Worker's remittance is a 600 billion dollar industry [3] that has been dominated by money transfer companies like Western Union [4] or MoneyGram [5]. But the delay in transfer and the cost of service, means that this is a rich area of disruption for Blockchain-based solutions. Indeed, companies from the United States, Hong-Kong, Singapore, Philippines, Mexico, and many other countries are trying to solve this problem by bringing cheap and quick solutions with Blockchain [6]. Even traditional companies like Western Union or banks like Santander are adopting Blockchain technology to keep themselves in the new world that Blockchain is ushering. Western Union has been partnering with Ripple [7] which offers enterprise payment solutions with Blockchain technology [8]. Santander UK, a traditional bank, has also been testing Blockchain for international payments by partnering with Ripple [9].

But it is not only about money; companies like Slock [10] are trying to use Blockchain and Internet of things (IoT) technology to create a Universal

Sharing Network (USN) where people can share or rent their products without a middleman. In this collaborative economy, the goal is to not let any asset remain underutilized. People can share any of their items for use of other people. Everything is recorded on a Blockchain, providing full autonomy. For example, if someone's apartment is vacant, it can be shared with proper compensation. Similarly, people can share Wi-Fi access, vehicles, and any other equipment. Industries can share their manufacturing tools for other projects. One of the other main goals in this venture is to have machine to machine communication, thereby removing needs for a human intervention. For example, a door to an apartment gets opened once the rent is paid, and things in the apartment can order their own repairs. Therefore, this will be an ecosystem of Blockchain and IoT, where autonomy will cut down costs, and bring in new revenue. One early use case of this ecosystem is where Slock partnered with European Energy company RWE to implement automatic charging stations [11]. In these stations, users can check in, pay the rent, and charge their cars. All transactions will be recorded in the Ethereum Blockchain, providing transparency for energy regulations to be enforced, and enabling new APIs to be developed for any future novel features.

Peer-to-peer energy sharing is being brought into the mainstream by companies like Power Ledger [12]. Distributed energy resources (DER) are a decentralized system of generating power. Traditionally, energy has been produced by a central powering house that uses traditional resources such as coal, gas, or nuclear energy. But they need to transfer the energy over very large distances. By contrast, DER systems are modular in nature, and as such, can be located near the area it will be serving. DER systems also use renewable energy sources such as solar or wind. Companies like Power Ledger are using Blockchain to incorporate into DER systems the capability of users to buy and sell their energy. They are using the Ethereum Blockchain [13], and introducing custom tokens to quantify the energy rate as well as the medium of buying and selling energy peer to peer [14]. Even third-party applications can be developed in this dynamic and scalable platform. LO3 Energy [15] is another company that is using a similar approach to build an ecosystem of affordable energy.

2.2 Insurance Industry

The insurance industry is one of the ripe places for disruption with its slow claims process, and over-reliance on manual work. Blockchain has huge potential to improve many aspects of this industry. Insurance claims take many week, and so much paperwork at the moment. A Blockchain-based system can have smart contracts which can have rules and regulations that can trigger some action once some conditions are met. A customer can upload documents of some insurance claim, and the Blockchain can then verify and instantly process that claim. By eliminating human intervention, this also saves logistics costs. Moreover, increasingly our devices are connected with Internet. IoT will surely improve over time. With so many devices connected to the cloud, a Blockchain can have constant flow of information from many types of devices, improving insurance industry's prediction power which is needed to predict risk. For example, in car insurance, the car can be connected to the cloud, and constantly feeding vital signs to the Blockchain. Once an incident happens, the company has full information to process the claim, which can be executed with a smart contract running on the Blockchain. Fraudulent claims are another aspect because of which insurances are expensive. With a shared Blockchain, data for each fraudulent claim can be inspected by all companies, reducing risks. Blockverify [16] is one such company that is trying to introduce anticounterfeit solutions with Blockchain into the supply chain management system. For health insurance, a Blockchain that can constantly aggregate health information from hospitals will help all companies gather data from a central, consistent database, providing speed for new customer signups, and claims processing. For health care data, privacy is of utmost importance. The strong security and privacy features of Blockchain can provide the desirable certainty in that regard. SimplyvitalHealth [17], a health care startup, is building a platform that takes advantage of machine learning and Blockchain. With Blockchain, it provides an immutable audit trail of sensitive data, and with machine learning, it provides forecasting of patient's health. This platform provides an interface for both hospitals and private insurance providers and implements an interoperable system where access to data is quick, and thereby decisions can be made quickly. Dynamis [18] is a company that is trying to use Blockchain to execute smart contracts that will provide unemployment insurance to people by integrating information from professional networks like LinkedIn. InsureETH [19] is another company that is offering travel delay insurance with an Ethereum smart contract. All claims are resolved automatically on the Blockchain as it keeps track of flight delay information.

2.3 Traditional Banking

Blockchain has the potential to cut costs from huge expenses incurred by traditional bank infrastructures. According to Santander Innoventures,

Blockchain can save traditional banks up to 20 billion dollars per year in infrastructure costs [20]. According to Accenture, Blockchain can save investment banks up to 12 billion dollars per year in infrastructure costs [21]. Companies like JPMorgan Chase are indeed investing in startups like Axoni [22] to work toward reducing their infrastructure costs [23]. One of the ways banks can save money is to automate the process of clearing and settlement for transactions. Clearing is the actual process of implementing a transaction after its initiation. Accenture predicted, investment banks can save up to 10 billion dollars by using Blockchain to support clearing [24]. Stock exchanges all over the world are testing the use of Blockchain to expedite settlement process, and remove administrative hurdles. Nasdaq has introduced its NasdaqLinqBlockchain on which a client issued his or her stock for the first time on a block chain to a private investor [25]. Using Blockchain 99% of the traditional 3-day processes can be decreased to just 10 min. The Australian stock exchange (ASX) is in fact working on using Blockchain to process all of its posttrade clearing [26]. ASX is working with US-based startup Digital Asset Holdings to replace its Clearing House Electronic SubregisterSystem (CHESS) with a Blockchain-based system. IBM and Japan's largest stock exchange Japan Exchange Group (JPX) are working on bringing Blockchain to low liquidity markets [27]. Korean Exchange together with a Korean Startup has introduced a Blockchain to trade equity shares of new startups [28]. Chile's Santiago exchange has partnered with IBM to use Blockchain to decrease processing times and administrative overhead [29]. Germany's Deutsche Börse, an exchange company, has worked on a prototype Blockchain system that can process settlements of securities [30]. Overall, many exchanges have internalized the benefits Blockchain technology can bring in.

Wholesale banking is another place where Blockchain can be used to save huge amount of cost. Wholesale banking refers to banking services between merchant banks and giant financial institutions. These are large transactions but incur astonishing amount of manual work. Blockchain can automate the process of serving these interests. Indeed, CLS Group [31], a leading provider for settlement services in the foreign exchange market, has started the Hyperledger Blockchain [32] for its payment services [33]. Traditionally, companies raise funds with angel investors, then with venture capitalists, and later, if successful, with Initial Public Offerings (IPOs). Blockchain offers a completely novel way to raise money by cutting the middlemen in this process. Many companies are now using Initial Coin Offerings (ICOs) [34] to raise money in a peer-to-peer way with no border restrictions. In 2016, companies have raised nearly 200 million dollars with ICOs [35]. Certainly, there is a lack of regulations for this type of crowdfunding at the moment. But as the technology matures, this will surely provide a fast way of raising money for globally supported projects. Blockchain can introduce much needed expediency and round the clock settlements in the equity market. Japanese investment firm SBI has partnered with Ripple to use Blockchain technology for payment services and settlements [36]. Overstock, an e-commerce company, for the first time issued some of its stocks that can be traded over Blockchain. It has also partnered with Keystone Capital Corporation to build its Blockchain platform for broker–dealers [37].

Reference data management is another sector Blockchain can streamline and bring consistency across institutions. Reference data refer to all data that are required to complete and settle transactions. Because of no automation, and inconsistency, institutions currently tend to keep their own reference data which further perpetuates the inconsistency. Blockchain can hold reference data in a dynamic way, as new data comes in, the information gets updated which in turn can be accessed immediately by all participating institutions. This will bring expediency and in turn incentivize business. Additionally, because of strong security guarantees, there is no chance of manipulating this information by malicious parties. In 2016, seven buy-side and sell-side firms have collaborated with Blockchain firms R3 and Axoni to simplify reference data handling process [38].

Another huge challenge for the traditional banks is the verification of their customers, and attributing proper identification to them. Because of strict regulations, to avoid money laundering or illegal financing, they have to maintain the integrity of the identity of their customers. This process costs a fortune in terms of infrastructure costs. Blockchain technology together with biometric identity systems can streamline this process by building shared digital identity storage that any finance institution can use. Because of security guarantees from cryptography, this data will also be protected. Accenture together with Microsoft is indeed working on such a project to provide identification properties to almost one-sixth of the population of the world who do not have such identification papers, nullifying their existence [39]. R3 [40], a startup, that is working on building a Blockchain-based operating system for banks, is also working on solving the fragmented identification system currently in existence. India's largest stock exchange, National Stock Exchange, with the technology of a startup has conducted a trial for a Blockchain-based Know-Your-Customer (KYC) data trial [41].

2.4 Microcredit for Poverty Alleviation

Microcredit is the established system of providing impoverished people with small loans with which they may start some small business and climb out of poverty. Microcredit has been used as a major weapon against poverty. Until 2009, approximately 74 million held a total microcredit of 38 billion dollars [42]. Yet, the efficacy of microcredit to alleviate poverty has often been questioned [43]. Research has also shown mixed results on the effectiveness of microcredit [44]. Because of the high logistics cost of conducting microcredit operations on the field, the interest rates have been traditionally very high. On average, the global interest rate for microcredits is at 37% [45]. This severely hampers the chance of impoverished person to get out of poverty and, in many cases, traps them into vicious cycle of loan and interest. Another problem in this area is the verification of identification of the borrowers, and determining their credit risks. As such, microcredit organizations tend to keep themselves at the safe side and deny many people who would otherwise be tremendously energetic entrepreneurs. Several companies are trying to use Blockchain to cut the middlemen together with the logistical costs and provide microcredits at a much lower interest rate. MicroMoney [46] is one such company that is trying to bring the 2 billion people outside any banking network but has access to a smartphone into the fold. Their goal is to provide microcredits to people without any kind of credit score and without any collaterals. To determine their credit risk, they are using deep learning with around 10,000 features collected through their smartphone application and the social network data of the applicants [47]. Through this process, they are also trying to build credit history for these people which will reside in a Blockchain, and thus can be used by other financial institutions to approach this huge population. Salt is another company that tries to solve the measuring credit worthiness problem by altogether removing credit score checking, and using Blockchain assets as collaterals when lending [48]. Because the collaterals are on a public, transparent system, the logistics of verifying someone's credit worthiness is much easier, and this smooths the lending process. The interest rate is also determined by the parties involved in the lending and the borrowing. Another platform that can be used for lending is built by JibrelNetwork [49]. People can transform their money or any other financial asset into CryptoDepositReceipts (CryDR) which can be used for remittances, payments, building any kind of financial instruments, and many other purposes. All of these currencies and other kinds of assets reside in the Ethereum Blockchain as standard ERC20 tokens, which is the standard for smart contracts on Ethereum.

2.5 Humanitarian Work

Humanitarian works require handing over of money and assets between many entities. As such, corruption is always a risk, especially in countries with weak infrastructure. Blockchain can bring much needed transparency, and consistency of data between many parties in real time, cutting out middlemen, corruption, and bringing expediency into the aid distribution network. World Food Programme (WFP) has recently conducted a successful pilot study of using Blockchain for its cash assistance program in Pakistan [50]. Whenever someone is assisted, that data is added into a public Blockchain based on Ethereum using a smartphone interface. That data is matched with the entitlements to verify the authenticity of distributions. After the success of this study, they are running a much bigger study in the Azraq refugee camp, Jordan [51]. People in the camp can buy food from stores in the camp with their biometric data, and that record is automatically added into a Blockchain. This ensures privacy of their biometric data by not sharing with any third parties and also provides quick and updated record of all the transactions happening to WFP.

3. APPLICATIONS TO NONMONETIZED ECONOMY

3.1 Preliminaries

This section is based on the paper [52]. After the invention of Bitcoin, cryptocurrency is now an accepted part of financial assets. But, it is a pile of digital information and hard to paint a precise picture actually. Bitcoin indeed has given a new twist to our concept of money. Now it poses the following question to us: "What is money?" and "What is capital?" From the ancient ages, human beings have used things presenting a real picture, like stones, shells, and gold for trades, and recently they stared using digital codes for warrants. Where will we go next? Viewing the recent affairs of economics from a different perspective, we notice that the market of sharing economy is getting big and that trend suggests people are not disinclined to use personal belongings of somebody else.

The economic system in which goods or services are directly exchanged for other goods or services without a medium of exchange is called barter. Reconsidering carefully, all trades of barter are done in a decentralized manner, and currency itself is an authorized medium. To get rid of central mediums is the main concept of Blockchain, and thereby barter may be a reasonable way to enjoy economic activities in a modern style. Until the last century, it had been commonly and widely believed from ancient times that money was invented to solve inconvenience of barter [53,54]. However, many economists and cultural anthropologists today send a skeptical look at this plausible story of the relation between money and barter [55–57]. According to them, there are no clear evidences backing the correctness of the theory, claiming that there was a society whose economy was built upon barter. Of course, such claims again do not justify the view that there were no economy systems based on barter in past days. Regarding the history of money and barter, this seems a never-ending debate. What we aim at in this chapter is not to add another viewpoint toward this debate, but to give a solution to construct a real and global economic system which is truly based on barter. In this chapter we explain the barter system, called Mariana, proposed in Ref. [52] and discuss why Mariana can be compatible with capitalism. In the modern society, economic activity is commonly evaluated by monetary measures. For example, Gross Domestic Product (GDP), which measures the market value of all goods and serviced produced in a period of time, is one of the barometers commonly used to evaluate economic power of countries. However, it does not always reflect affluence of countries. Everyone who experienced in traveling abroad would notice this gap without looking up economic database. Giving qualitative evaluation to the degree of such a gap might be impossible, since money only represents one aspect of goods or services and the other aspects which are not evaluated by money are neglected in this monetized economy system today. Our solution toward this situation is to generalize capitalism today in such a way that products are evaluated by multidirectionally. With some reflection, capitalism is not necessarily on the axis of monetized economy. It looks so since the society today is too much dependent on monetized economy and, out of nowhere, a hierarchical world based on monetary assets has been established though money is nothing but a medium of exchange. Main purpose of our system is to endow this capitalist society with multiple axes on which the value of goods or services are projected.

3.2 Peer-to-Peer Global Barter Web

Here we explain our peer–to–peer system employed to Mariana. Its novelties can be summarized as follows:

1. It can work without proof of works.
2. Miners can dig out as much as they wish.

The first point is very different from any of the conventional Blockchain systems. Since barter basically goes through by two persons agreements, it does not require to form public consensus. Historically, a Blockchain was invented in terms of making a cryptocurrency system which exists on a peer-to-peer network and transactions do not go through any third party. In this system reaching a public agreement is very important to decide price of currency and accept transactions officially. However, this process causes another problems. Fifty-one percent of the participants are required to approve transactions to add and renew a block, but it happens that the majority might malignantly accept incorrect transaction records. Solving this problem is still an open problem. Instead of finding a complete solution to this issue, we change the problem setting and construct a peer-to-peer system free from the 51% problem. Second, the mining procedure in Mariana is also very different from the conventional cryptocurrencies, most of which have the maximal limit of the amount of deposit, whereas, in Mariana, everyone can bring whatever they want to trade into the market. This procedure corresponds to mining. More precisely, it can be described as follows. The mining process resembles to an auction in the sense that the wining person who achieved an agreement for trading can add a block to the original chain, where each block includes personal belongings of an owner. Suppose four persons Alice, Bob, Charlotte, and David are in the market and Alice and Bob want to get a certain item I_C belonging to Charlotte:

1. Alice and Bob send trade requests by themselves and several candidates for trade-off to Charlotte.
2. Charlotte checks the lists of candidates offered by Alice and Bob.
3a. Suppose Charlotte finds a desirable item I_B belonging to Bob. Then she renews the belonging lists and wish lists of her and Bod, adds a block to the chain, and opens the lists to public.
3b. Suppose there are no items Charlotte desires to trade. Then, the lists remain the same as before.
4. Alice, Bob, and David check that Charlotte has updated the lists correctly (Fig. 1).

In comparison with conventional Blockchain systems, novelties of this system are as follows. First of all, an observer (David) is not involved into a business among Alice, Bob, and Charlotte and he respects their agreements; thereby there is no need of the general proof of work. This is because that the value of items or services may vary from person to person and barter is based on personal agreements, whereas the value of money or cryptocurrency must be formed by public consensus. Hence the Bitcoin protocol requires persons who are irrelevant to transaction to join in proof of work.

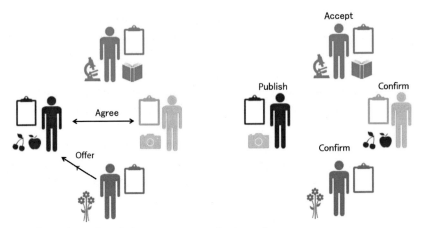

Fig. 1 Procedure of trade in a peer-to-peer barter web.

In this way, Mariana never suffers from the 51% problem as a matter of course and it rather encourages people in market to trade their products based on their own values. Here we mean by the 51% problem by that major cryptocurrency protocols as represented by Bitcoin approve transactions if more than 51% of the attendance accepts the results of computation. However, this process of consensus building can cause a problem if a large number of malicious traders participate in the mining. Second, mining process of Mariana is also different from that of the usual cryptocurrency systems, in which the total amount of coins is limited and they often experience excessive price volatilities due to trades for speculation purposes. On the other hand, mining in Mariana is completed by cultivating user's resources and offering whatever they recognize its value. Thereby it is internal and external mining.

It is surely possible to make a similar system in a centralized manner; however, one of the incentives to use a Blockchain-like way is that it may contribute to reduce costs of managing a central server by dedicating necessary calculations to users. Moreover, since Blockchain is an open source technology, it will be helpful for other users to join in the market freely and develop the system, which will allows one to provide various services as what the cryptocurrency trades at many different exchanges are.

3.3 Logistics for the Peer-to-Peer System

Now let us reconsider a commodity distribution web mostly employed for the major systems today. In the modern society, partly centralized networks,

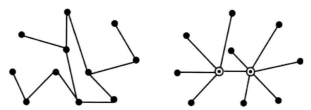

Fig. 2 Random network and scale-free network.

called a scale-free network (right-hand side of Fig. 2), have been preferred to handle medium- or large-sized systems effectively.

In comparison with a random network, a scale-free network has an optimized network apparently; however, it becomes vulnerable once accesses flock to a hub. Indeed, the logistics industry in Japan suffers from this problem. When the number of total packages is very large, access to each hub is condense and such a superloaded hub gets difficult to manage distribution of packages successfully. As a consequence, a random network scheme is marred to system failure. This situation is similar to the traffic jam which happens around junctions in highway. What we want to stress is that this is an inescapable problem of this type of traditional distribution web. Therefore in order to stave off potential system trouble, it is required to manage a flood of packages to deriver in an alternative way so as to reduce burden of each hub. The following examples give crucial evident facts of the situation we discussed above with a good amount of data. The report [58] describes the actual status of the logistics industry in Japan and according to it, Yamato Transport leads the industry and their total deriver volume has reached 1.8 billion parcels. Moreover, the major customer of Yamato Transport is Amazon Japan and it is about 17% of their total deriver volume. Yamato Transport has taken on all delivery assignment by Amazon Japan. This is because Sagawa Transport, the second biggest transportation company in this industry, terminated their trade agreements with Amazon Japan in 2013 since their core business was corporate delivery services and they did not need to have a detailed physical distribution network. The other transportation companies today are not in charge of shipping packages from Amazon Japan because of their logistics systems and strategies. However, there is a rampant problem among the logistics industry in Japan. According to the survey report on redelivery by MLIT (Ministry of Land, Infrastructure, Transport and Tourism) in 2015, it is reported that the number of delivered packages occupies about 20% of the total delivery package and it amounts to 0.81 million packages [59]. In order to deal with the increase in the number of shipping packages, Yamato Transport and Amazon reached to an agreement that Amazon pays 40% more for shipping in Japan. Existence of those redelivery packages leads to functionally deficient of the logistics systems. In addition to that, it is also important to reduce the amount of packages which center around each local distribution depot so that the logistics system recovers its efficiency.

Mariana addresses the urgent issue discussed in the previous subsection by employing both of a random network and a scale-free network. We present a new logistics illustrated in Fig. 3. We explain that this logistics network should be compatible with the concept of barter and the modern retail business. Our proposal for a next type of logistics network consists of the following two concepts.

First of all, we aim at recovering mobility of a distribution network and improving its capability. For this purpose, it will be mostly effective to reduce the number of packages flowing into each hub. The most simple way is to increase the number of hubs. Placing delivery lockers at stations and forming a strategic business alliance with a franchising company, so that customers can receive or send products at franchisees, are typical examples of possible and effective solutions. Indeed, several convenience stores in Japan and shipping companies offer this service. However, this is still not enough to solve the problem since adding hubs into a scale-free network does not change the fundamental structure of the network. In the commodity distribution network today, packages mostly flow in one direction. For example, ordinary customers often enjoy online shopping to buy something and products are distributed from stores to customers, whereas packages rarely flow from customers to stores. Readers may take it for granted; however, the authors consider that turning a watchful eye toward this situation would give us a hint to solve the shipping problem. It would be not too much to say that this kind of one directional flow of products is one of

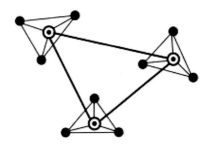

Fig. 3 Logistics for a peer-to-peer network.

the causes of mass production and mass consumption in the modern society. However, most of the products owned by customers are not used so often and consequently such unused items are destined to be stored at their houses. Namely, houses are warehouses. So if people reutilize them, multidirectional flow of items should be realized and more environment friendly business network would be established. In Mariana, each hub is expected to play more or less a similar role of exchanges dealing cryptocurrencies. In other words, it offers people living closely to trade products directly without going through a third party (they may use hubs if indirect trades are preferred), and large hubs are basically used for remote trades.

3.4 Postcapitalism

Here we consider Mariana's place within the history of monetized economy and nonmonetized economy in order to clarify the novelty of the economic system we are suggesting in this chapter.

3.4.1 Era of Central Bank

Before money was invented, it is believed that people exchanged things and services they want for something equivalent in return in a system, called barter system. The beginning of bartering is sometimes said to be around 6000 BC and adopted by Phoenicians and developed by Babylonians in a sophisticated way. Though any concrete evidence showing the exact era when currency was invented and used is not known today, ancient people somehow started using currency and merchants who played the role of banks today appeared with an increase in the use of money. It is known that the central bank systems today have their origin in the late 17th century. The main roles of a central bank are to issue currency and to maintain financial stability. To achieve these purposes, it carries out monetary control and ensures smooth settlement of funds among banks and other financial institutions. Central banks seem to be in a turning point due to the enormous amount of money supply and the huge increase in debt worldwide as shown in Figs. 4 and 5.

Fig. 4 shows changes in the current-year broad money, which is the most inclusive method of calculating a given money supply that generally includes demand deposits at commercial banks and any money in addition to physical money such as currency and coins. This estimation is based on International Monetary Fund, International Financial Statistics and data files, and World Bank and OECD GDP [60]. According to it, total broad money in the world has been monotonically increasing from a macroscopic viewpoint

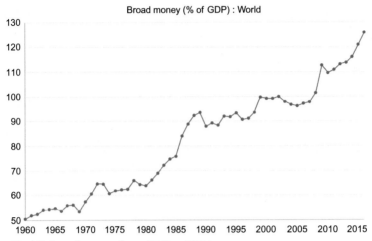

Fig. 4 World's broad money from 1950 to 2015.

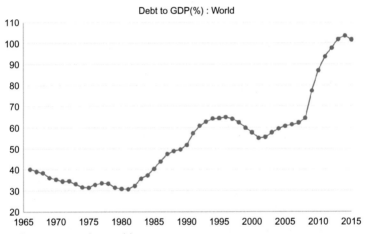

Fig. 5 Debt to GDP in the world.

and especially broad money has shown a rapid increase from 2013, when cryptocurrency as represented by Bitcoin attracted general interests of investors most successfully. We will revisit this theme in the next step.

Fig. 5 shows current-year world debt to world GDP, and according to it, the debt situations highly depend on each period. One of the most interesting things is that in comparison to broad money in the figure showing changes in the broad money, on one hand, the debt had been decreased and the broad money had been increased from the 1960s to the 1970s, and on the other hand the debt has been increased and the broad money also has been increased from the 2000s (the similar situations can be seen in the period of the 1980s to the 1990s). These facts suggest that it is difficult to cope with modern economical problems only by the central and commercial banking system.

3.4.2 Postcapitalism

Modern information science and new technologies such as the Internet and Blockchain enable us to find a trading partner with reduced cost because of the access to huge amount of information and higher possibility of coincidence of wants. Capitalism will evolve to a next stage with the concept of barter with a challenge to put money, which was in a position of inflation to put back into a position where it has to be. Capitalism today is too much dependent on money, so we may begin with recalling the role of money in order to discuss the advanced capitalism. The question we should reconsider is what money does measure. Almost all of the economic activities are evaluated by monetized indices and hence economical profit is understood to be equivalent to the amount of earned money. But definition of the capitalism does not require such a single framework. Under ordinary circumstances, capitalism should be based on the genuine and general capital. Hence everyone with or without money is supposed to be able to join in the market. Financial affluence does not always link to real abundance in ordinal activities and lives of people; however, what the capitalism has become today seems to send a wrong message to human beings in future.

Therefore our solution to this situation is to develop the modern capitalism framework so that it accepts more general indices to evaluate economic activity. To do so, we first propose to generalize a role of money in such a way that it should be treated as just an item to exchange. That is to say, money will still have a value as a meditation, but other items or products will be put to the same level of the money and offer people equivalent chances to use them for exchanging. For example, in our system people may be allowed to "pay" housing tax by electrical power generated by solar panels equipped with their houses.

3.5 Difference From Sharing Economy

These days, the concept of a sharing economy is blooming all over the world. Its market size is estimated to grow from 14 billion dollars in 2014 to 225 billion dollars in 2025. A sharing economy is an economic model in which individuals are able to borrow or rent assets owned by someone else.

Readers might find Mariana similar in many points. But there are several fundamental differences from it. The most obvious thing is that in a sharing economy, we still need money to get in this community [61], but in Mariana, we do not use money and we do possess things we want by exchanging our belongings. In addition to that, "sharing" offered by most of the conventional web services available today, which play crucial roles in a sharing economy, is established in the basis of on one-way communication. For example, when one wants to rent a room which is not used tentatively by the original owner, he/she books the room using an app and pay money to the owner. In the process of this business, customers can choose rooms but owners cannot choose customers. Moreover, customers have a good number of opportunities to choose rooms from various examples, whereas owners are allowed to receive only a single medium (usually money) as a reword.

What the authors want to establish in this chapter is to provide an alternative economic model based on our novel Blockchain architecture in such a way that it is compatible with the traditional sharing economy and provides all users equal opportunities to negotiate in the market so that they can enjoy their economic activities in preferred ways. As frequently mentioned in this paper, the modern society faces excessive money supply and suffers from ill-matched debt against the overall GDP. Since barter is very old decentralized system existing before the birth of Blockchain, it will be quite natural to revisit and shed a new light on its possibility.

4. SUMMARY AND FUTURE RESEARCH DIRECTIONS

In this chapter, we reviewed the latest and greatest applications and ventures of Blockchain in the financial sector. These applications have immense potential to redefine humanity, and move us toward a frictionless system where majority of logistical delays and red tape can be cut down. And also we explained a novel peer-to-peer system which enables people to engage in a new economic activity integrating barter, sharing economy, and capitalism. This system is new as a Blockchain because of the following two points:

1. It works without proof of work.
2. It works by internal and external mining.

Here we mean that it works without the proof of work by that it does not require building public consensus and the minimum number of people for accepting transactions is only two. In addition to that the mining procedure is totally based on cultivating resources of users and it encourages them to be as creative as possible. This is quite different from the mining proposed in

Bitcoin, whose total amount of coins is limited and there is no room for users to produce coins but to find embedded coins with huge machine power as if they were sandgropers. One aim of Mariana is to shed new light on products against this old-fashioned way to evaluate value of products. Namely, Mariana's philosophy is based on respecting features of products from various points of view. This would lead to respect diversity of user's personalities since belongings more or less represent characters of owners. Furthermore the proposed economic activity will be placed at a new position of post-capitalism and postmonetized economy, though it does not aim to exclude money itself from society. Rather it integrates all of good points of barter, monetized money economy, and capitalism. In some sense Mariana is a generalization of monetized economy since it allows one to pay by anything, including money, as far as they reach an agreement. Moreover, it is a new type of capitalism since it may encourage people in market to create something having a unique value. In this system things which may be difficult to be priced by a single medium, money, will be evaluated in a more fair way, and thereby creators can accept better rewards. Moreover, this system would be the first global barter system over the course of recorded human history. Concept of barter is nothing new; however, it would be clear that there had been no ways to establish the barter system in such a way that every people can join trading regardless of inhabited areas. The authors believe that this system will take the modern society to a new stage of economic regime and will have an affinity for our modern society which respects diversity and cultural identity. In order to promote this affair, we consider it is important to introduce an alternative way to estimate value of products not only by money. Mariana is one of the solutions to show our attitude. We plan to conduct several social experiments in several cities in the world and relevant estimations will be reported in the near future.

KEY TERMINOLOGY AND DEFINITIONS

Barter It is the simplest way to exchange goods or services for other goods or services without using a medium of exchange. Barter has been thought to be a major way to build a regional economic community in ancient times.

Nonmonetized economy It includes the civic services or social services that are indispensable for the sustainability of a functioning society yet cannot be quantified in terms of monetary values.

Monetized economy It is an economy system in which goods or services for other goods or services by means of money. The modern monetized economy system has established only in the 18th century, when the central banks played the fundamental roles in the market. However, this system has experienced several crises because of changes in the value of money and erosion of trust of the government or banks issuing money.

REFERENCES

[1] Send money like a text. No fees. No borders. No one else is doing this. Circle | On a Mission to Create a More Inclusive Global Economy. https://www.circle.com/en.

[2] Digital Money Solutions on Distributed Ledger Technology. Billon. https://billongroup.com/technology/.

[3] Remittance, Wikipedia. https://en.wikipedia.org/wiki/Remittance.

[4] Money Transfer | Global Money Transfer | WU United States. https://www.westernunion.com/us/en/home.html.

[5] MoneyGram website. https://secure.moneygram.com/.

[6] 11 Money Transfer Companies Using Blockchain Technology. https://gomedici.com/11-money-transfer-companies-using-Blockchain-technology-2/.

[7] Ripple website. https://ripple.com/.

[8] Western Union Says It's Testing Transactions With Ripple. https://www.bloomberg.com/news/articles/2018-02-13/western-union-says-it-s-testing-transactions-with-ripple.

[9] Santander Becomes First U.K. Bank to Introduce Blockchain Technology for International Payments. https://bitcoinmagazine.com/articles/santander-becomes-first-u-k-bank-to-introduce-Blockchain-technology-for-international-payments-1464795902/.

[10] Slock Website. https://slock.it/.

[11] Forget About Oil, Your Next Car Will Run on Ethereum! https://www.financemagnates.com/cryptocurrency/innovation/forget-about-oil-your-next-car-will-run-on-ethereum/.

[12] Power Ledger Website. https://powerledger.io.

[13] Ethereum Blockchain Website. https://www.ethereum.org/.

[14] Power Ledger White Paper. https://powerledger.io/media/Power-Ledger-Whitepaper-v8.pdf.

[15] LO3 Energy, Empowering Communities Through Localized Energy Solutions. https://lo3energy.com/innovations/.

[16] Blockverify Website. http://www.blockverify.io/.

[17] Simply Vital Health Website. https://www.simplyvitalhealth.com/.

[18] Dynamis Website. http://dynamisapp.com/.

[19] InsureETH Website. http://insureth.mkvd.net/.

[20] Santander: Blockchain Tech Can Save Banks $20 Billion a Year. https://www.coindesk.com/santander-Blockchain-tech-can-save-banks-20-billion-a-year/.

[21] Blockchain Could Save Investment Banks Up to $12 Billion a Year: Accenture. https://www.reuters.com/article/us-banks-Blockchain-accenture/Blockchain-could-save-investment-banks-up-to-12-billion-a-year-accenture-idUSKBN1511OU.

[22] Axoni Website. https://axoni.com.

[23] Goldman, JPMorgan to Invest in Blockchain Startup Axoni: Sources. https://www.reuters.com/article/us-axoni-Blockchain/goldman-jpmorgan-to-invest-in-Blockchain-startup-axoni-sources-idUSKBN149073.

[24] Banking on Blockchain—A Value Analysis, Accenture Consulting. https://www.accenture.com/t20170120T074124Z__w__/us-en/_acnmedia/Accenture/Conversion-Assets/DotCom/Documents/Global/PDF/Consulting/Accenture-Banking-on-Blockchain.pdf#zoom=50.

[25] NasdaqLinq Issues Shares With Blockchain Technology. http://allcoinsnews.com/2016/01/04/nasdaq-linq-issues-first-shares-with-Blockchain-technology/.

[26] Five Ways Banks Are Using Blockchain, Financial Times. https://www.ft.com/content/615b3bd8-97a9-11e7-a652-cde3f882dd7b.

[27] IBM and Japan's Largest Stock Exchange to Test Blockchain for Trading Environments. https://www-03.ibm.com/press/us/en/pressrelease/49088.wss.

[28] Korea Exchange Opens Its KRX Startup Market Exchange With Blocko'sBlockchain Technology. https://www.businesswire.com/news/home/20161116005521/en/Korea-Exchange-Opens-KRX-Startup-Market-Exchange.

[29] IBM and Chile's Santiago Exchange to Deliver World's First Securities Lending Blockchain Solution. http://www-03.ibm.com/press/us/en/pressrelease/52407.wss.

[30] Joint Deutsche Bundesbank and Deutsche BörseBlockchain Prototype. https://www.bundesbank.de/Redaktion/EN/Downloads/Press/Pressenotizen/2016/2016_11_28_Blockchain_prototype.pdf.

[31] CLS Group Website. https://www.cls-group.com/about-us.

[32] HyperLedgerBlockchain Website. https://www.hyperledger.org/.

[33] Currency Settlement Service CLS Reveals Big Blockchain Ambitions. https://www.coindesk.com/cls-to-develop-Blockchain-payment-service-on-ibm-fabric/.

[34] Initial Coin Offering, Wikipedia. https://en.wikipedia.org/wiki/Initial_coin_offering.

[35] Can You Trust Crypto-token Crowdfunding? https://techcrunch.com/2017/02/12/can-you-trust-crypto-token-crowdfunding/

[36] SBI Ripple Asia Announces Japanese Bank Consortium. https://ripple.com/insights/sbi-ripple-asia-announces-japanese-bank-consortium/.

[37] Overstock: Broker-Dealer Deal Will Open Blockchain Floodgates. https://www.coindesk.com/tos-first-broker-dealer-could-open-Blockchain-securities-floodgates/.

[38] Press Release: R3 and Axoni Explore the Use of Distributed Ledger Technology to Reduce Risk in Reference Data Management With and Buy and Sell Side Financial Services Firms. https://www.r3cev.com/press/2016/9/20/press-release-r3-and-axoni-explore-the-use-of-distributed-ledger-technology-to-reduce-risk-in-reference-data-management-with-buy-and-sell-side-financial-services-firms.

[39] Accenture, Microsoft Create Blockchain Solution to Support ID2020. https://newsroom.accenture.com/news/accenture-microsoft-create-Blockchain-solution-to-support-id2020.htm.

[40] R3 Website. https://www.r3.com/.

[41] India's Biggest Stock Exchange is Testing Blockchain KYC. https://www.coindesk.com/india-stock-exchange-Blockchain-kyc/.

[42] Microcredit, Wikipedia. https://en.wikipedia.org/wiki/Microcredit.

[43] Microloans Don't Solve Poverty. https://fivethirtyeight.com/features/microloans-dont-solve-poverty/.

[44] Measuring the Impact of Microfinance in Hyderabad, India—Abhijit Banerjee, Esther Duflo, Rachel Glennerster, Cynthia Kinnan. https://www.povertyactionlab.org/evaluation/measuring-impact-microfinance-hyderabad-india.

[45] Banks Making Big Profits From Tiny Loans. http://www.nytimes.com/2010/04/14/world/14microfinance.html.

[46] MicroMoney Website. https://www.micromoney.io/.

[47] MicroMoney White Paper. https://www.micromoney.io/MicroMoney_whitepaper_ENG.pdf.

[48] Salt Company Website. https://www.saltlending.com/.

[49] Jibrel Network Website. https://jibrel.network/.

[50] What Is 'Blockchain' and How Is It Connected to Fighting Hunger? https://insight.wfp.org/what-is-Blockchain-and-how-is-it-connected-to-fighting-hunger-7f1b42da9fe.

[51] Blockchain Against Hunger: Harnessing Technology in Support of Syrian Refugees. https://www.wfp.org/news/news-release/Blockchain-against-hunger-harnessing-technology-support-syrian-refugees.

[52] K. Ikeda, Barter as a Peer-to-Peer Economic Network: Capitalism in the Blockchain Era, to appear, 2018.

[53] J. Locke, Locke: Two Treatises of Government Student Edition, Cambridge University Press, 1988.

[54] A. Smith, R. McCulloch, An Inquiry Into the Nature and Causes of the Wealth of Nations, A. and C. Black and W. Tait, 1838.

[55] G. Dalton, Barter, J. Econ. Issues 16 (1) (1982) 181–190.

[56] C.P. Kindleberger, A Financial History of Western Europe: Routledge, 2015.

[57] D. Graeber, Debt: The First Five Thousand Years, Eurozine, 2009.

[58] Ministry of Land, Infrastructure, Transport and Tourism, White Paper on Land, Infrastructure, Transport and Tourism in Japan, 2015 (Online). http://www.mlit.go.jp/common/001157851.pdf, 2015.

[59] Yamato-Holdings. Delivery Market Trends. [Online]. http://www.yamato-hd.co.jp/english/investors/market.html

[60] World Bank. Broad Money. [Online]. https://data.worldbank.org/indicator/FM.LBL.BMNY.GD.ZS

[61] C.J. Martin, The sharing economy: a pathway to sustainability or a nightmarish form of neoliberal capitalism? Ecol. Econ. 121 (2016) 149–159.

ABOUT THE AUTHORS

Kazuki Ikeda is a PhD student at Osaka University, Japan, Japan. He was selected to an Honorable Graduate Student of Graduate School of Science, Osaka University (2015–20). He was selected to a creator of MITO Advanced Program, which is a part of Exploratory IT Human Resources Project run by Information-Technology-Promotion-Agency, Japan (IPA) (2017).

His research interests include quantum physics, mathematical physics, information science (both classical and quantum), and artificial intelligence. His research papers are published in many conferences, workshops, and International Journals of repute including IEEE.

Md-Nafiz Hamid is a PhD student at Iowa State University, USA where he received a graduate fellowship at the Bioinformatics and Computational Biology program. His research involves applying machine learning and deep learning techniques to solve problems in the antibiotic resistance domain.

CHAPTER FIVE

Blockchain for Business: Next-Generation Enterprise Artificial Intelligence Systems

Melanie Swan
Purdue University, West Lafayette, IN, United States

Contents

Advances in Computers, Volume 111
ISSN 0065-2458
https://doi.org/10.1016/bs.adcom.2018.03.013

Abstract

This chapter discusses Blockchain distributed ledgers in the context of public and private Blockchains, enterprise Blockchain deployments, and the role of Blockchains in next-generation artificial intelligence systems, notably deep learning Blockchains. Blockchain technology is a software protocol for the secure transfer of unique instances of value (e.g., money, property, contracts, and identity credentials) via the Internet without requiring a third-party intermediary such as a bank or government. Public Blockchains such as Bitcoin and Ethereum are *trustless* (human counterparties and intermediaries do not need to be trusted, just the software) and *permissionless* (open use), whereas private Blockchains are *trusted* and *permissioned*. Enterprise Blockchains are private (trusted, not trustless) immutable decentralized ledgers, with varying methods of reaching consensus (validating and recording transactions). Four enterprise systems are examined: R3's Corda, Ethereum Quorum, Hyperledger Fabric, and Ripple. New business analytics and data science methods are needed such as next-generation artificial intelligence solutions in the form of deep learning algorithms together with Blockchains. The hidden benefit of Blockchain for data analytics is its role in creating "clean data": validated, trustable, interoperable, and standardized data. Business Blockchains may develop across industry supply chains with shared business logic and processes, and shared financial ledgers. Payment channels and smart contract asset pledging may allow net settlement across supply chains and reduce debt and working capital obligations. Specific use cases are considered in global automotive supply chains, healthcare, digital identity credentialing, higher education, digital collectibles (CryptoKitties), and asset tokens (Primalbase).

1. INTRODUCTION

Blockchain technology was inaugurated in 2008 with Nakamoto's white paper "Bitcoin: A Peer-to-Peer Electronic Cash System" [1]. The first Blockchain is Bitcoin. The original purpose and first use case of Blockchain technology were the implementation of the cryptocurrency, Bitcoin. Beyond their use in the economic domain, Bitcoin and Blockchain technology as articulated by Nakamoto solve an important computer science problem that had been a barrier to having a functional digital monetary system for years: the double-spending problem. The double-spending problem is that money should only be spent once, unlike a file, which might be copied arbitrarily many times. The first Bitcoin transactions occurred in January 2009. Almost 10 years later in 2018, distributed ledger technology implementations are well underway in a variety of areas.

Blockchain distributed ledgers are one of the fastest-adopted technologies in history. UK-based market research firm Juniper said that (as of July 2017) 57% of large corporations worldwide (companies with more than 20,000 employees) were either actively considering or in the process of deploying Blockchain [2]. Two-thirds of these firms said they expected to ingrate Blockchain distributed ledgers into their enterprise IT systems by the end of 2018. Corporate, industrial, and government interest in Blockchain technologies is high because applications extend well beyond the domain of cryptocurrencies. There are four main application classes for Blockchain technology: (1) monetary assets (currency, payments, remittance, finance, securities, and financial instruments), (2) property (land, real estate, and auto title registries), (3) contracts (business agreements, licensing, registration, wills and trusts, partnership agreements, and IP registration), and (4) identity credentials (passport, visa, driver's license, and birth registries).

1.1 Public and Private Blockchains

There are two kinds of Blockchain networks: public and private. Public Blockchains are *trustless* (meaning that users need to trust the software, but not human counterparties such as banks or governments) and *permissionless* (meaning that anyone can join and use the network). Private Blockchains on the other hand are *trusted* (meaning that users need to know and trust the human-based network provider and counterparties (individuals, companies, and governments)), and *permissioned* (meaning that users are known and credentialed).

Public Blockchain networks such as Bitcoin, Ethereum, Litecoin, and any of the ~1500 cryptocurrencies listed at CoinMarketCap, and the 1127 (March 2018) Ethereum token projects listed at https://www. stateofthedapps.com/ are public. This means that anyone can download a wallet, obtain the native cryptotoken, and start conducting transactions on the network. Identity is pseudonymous (meaning using an address hash (a 32-character alphanumeric sequence to transfer funds)) or fully confidential (using ZKPs (zero-knowledge proofs) or another technology feature to mask sender and receiver address and the amount sent). Many Blockchains are starting to offer confidential transactions, such as Monero, Zcash, and Ethereum. The implication is that true Internet privacy might be possible that lets the user prove certain information without revealing it. This could include someone proving that his or her age is over 18 without revealing the date of birth, or proving that there is enough money in the bank for a financial transaction without revealing the total balance or other details. The properties of public Blockchain networks are useful to understand because enterprise networks draw from the same kinds of decentralized architectures, concepts, design principles, and tools.

Private Blockchain networks comprise the majority of corporate, industrial, and government Blockchain projects. In private Blockchains, users are known, have login credentials issued by a trusted authority, and are likely assigned to a certain limited scope of activities on the network. Private Blockchain network users may have specific kinds of ledger writing access. Various other parties may have different private read-only views of the data (for example, compliance officers and regulators). It is important to understand the parameters and operation of any private network a user might consider joining. For example, if you are a supplier thinking of using an automobile manufacturer-sponsored supply chain Blockchain, it is necessary to trust how the sponsoring party is administering the private Blockchain.

2. PUBLIC BLOCKCHAINS: BITCOIN AND BLOCKCHAIN TECHNOLOGY

2.1 What Is Blockchain Technology?

Blockchain technology is a software protocol which enables the secure transfer of money, assets, and information via the Internet, without the need for a third-party intermediary such as banks or other financial institutions [3]. Transactions are validated, executed, and recorded chronologically in an append-only and tamper-resistant database, where they remain always available on the Internet 24/7 for on-demand lookup and verification. Blockchain technology is what underpins applications such as the Bitcoin cryptocurrency. Bitcoin was the first and perhaps most obvious application of Blockchain technology, as a decentralized payment system allowing for the real-time transfer of digital currency, at any time and to anyone in the world. Just as SMTP (simple mail transfer protocol provides the underlying software protocol by which Internet users can send an email to each other in a seamless and interoperable way, regardless of their email provider, likewise, the Bitcoin protocol allows people to seamlessly transfer money with one another, regardless of their bank. Money (and value transfer and secure information confirmation) become an IP designation, an Internet content type; a type of Internet traffic that is routed on the public network as any other; the concept is "money over IP."

With Bitcoin, money can be transferred from one continent to another, at very low costs and in a matter of seconds, instead of waiting days or weeks, and paying high commissions, as is the case with current international money transfers and remittance solutions. But money transfer is just one application enabled by Blockchain technology. The same technology also provides the means to record and transmit digital goods over the Internet, while ensuring that these goods cannot be copied or multiplied (thereby addressing the double-spending problem that has been an issue with digital currencies). Indeed, once they have been digitized as "smart assets," the recording, search, purchase, sale, tracking, and logging of resources can be coordinated with a much higher degree of automation, speed, trackability, and assurance [4]. A Blockchain can be used, for instance, as a digital registry to record, transfer, and verify asset ownership (home, auto, stocks, bonds, mortgages, and insurance) as well as to preserve the integrity and authenticity of sensitive documents or records (e.g., passports, visas, driver's licenses, birth and death certificates, voter registration, contracts, wills, patents, and medical records). Eventually, as governments, corporations, and start-up companies work toward the implementation of real-time payments, and digital registration systems for the transfer and verification of digital assets and legal documents, a variety of legal, financial, and governmental services could be reengineered and readjusted for the Internet era. One implication of transferring value with Blockchain-based smart networks instead of relying on human-based institutions is that the traditional intermediaries responsible for verifying and validating transactions may become obsolete. As a result, the institutional structure of society could shift to one that is computationally based and thus has a diminished need for human-operated brick-and-mortar institutions.

2.2 Blockchain and Business

Blockchain distributed ledgers are the latest and most profound form of Fintech (financial technology). They are an enterprise technology that ideally fits into existing IT infrastructures fairly seamlessly, in a plug-and-play manner with web services, middleware, and backend software. In doing so, Blockchain distributed ledgers are fundamentally creating a new way of conducting the economic and information transactions that are central to business operations, by using secure communications networks. As mentioned earlier, the concept is "money over IP" in the sense of designating money (unique instances of value) as an Internet protocol, analogous to any other traffic type. This means that the Internet has instructions about how to route value-related traffic when it comes onto the network. The key properties of Blockchain networks are that they are simultaneously private and transparent, distributed, immutable, and consensus based. It is novel to have privacy and transparency at the same time, readily usable in a standard and widely accessible network technology. Previously it was possible but expensive to set up point-to-point VPN connections (virtual private networks) and EDI (electronic data interchange) among parties. In Blockchain networks, not only are transactions private, but there is also transparency in the sense of knowing that other parties are complying with the same rules. Technical features such as hashing and ZKPs validate information without revealing its contents. Thus, another party's activity can be confirmed, without having to disclose the details. Any network-based activity follows the same validation protocols, and this enforces good player behavior. The benefit is that parties retain privacy while signaling their participation in trustful rule-based societal practices [5].

Some of the kinds of value transfer that take place using Blockchain networks involve monetary assets (including cryptocurrencies), property assets (digitally instantiated assets), business contracts (purchase orders, sales agreements, proof of goods delivery), and identity credentials. For inventory and transfer, assets are registered in a distributed ledger system. This means that they exist in a prestaged, validated, and standardized format, ready for on-demand confirmation and transfer. The implication is that asset transfer can be instantaneous. The further implication is that a wide-spread business and consumer practice of asset logging and transfer with Blockchains creates a new method of storing and transferring value that is inherently more trustful. Trust is created in that corporate and personal assets registered to Blockchains means that value is prevalidated and stored in a more usable way, which can be readily mobilized on demand. No asset can exist in the system without having been preconfirmed, so there is trust in the value of not only individual assets but an entire system in which all assets have been prevalidated. The societal practice of how we store and mobilize value is changing and becoming more trustful with Blockchain economic networks [6].

These two factors, the real-time transactability of all assets on Blockchain systems and the real-time information climate about network activity (magnitude and type, not specific details), work together to produce a new class of trust which we term *algorithmic trust*. Algorithmic trust is not the computerized instantiation of trust between human parties, but a new concept of trust that emanates from the use of the computational system as a replacement for traditional mechanisms of governance, control, and interaction. Algorithmic trust is an information affordance that is a new form of social capital, and is observable as a resource in Blockchain networks.

2.2.1 Distinction Between Blockchain Technology and Distributed Ledger Technology

Although the terms Blockchain and distributed ledgers are often used synonymously, there is a distinction. The general form of the technology is distributed ledgers, and the specific form of the technology is Blockchain technology (literally blocks chained together by cryptographic hashes). The general form of the technology, distributed ledgers, can be defined as follows. A ledger is a file that keeps track of who owns what. A distributed ledger is defined by there being (1) a shared transaction database among network members, which is (2) updated by a consensus algorithm, in which (3) records are timestamped with a unique cryptographic signature, appearing (4) in a tamper-proof auditable history all transactions. Blockchain, then is a specific form of distributed ledger technology with an additional feature, the (5) sequential updating of database records per chained cryptographic hash-linked blocks. In the Blockchain economy (e.g., the implementation of distributed ledger technologies), it is important to distinguish between their different feature sets. One reason is that private "Blockchains" deployed in enterprise use are often not actually Blockchains per this definition. The main distinction is regarding the properties of decentralization and consensus algorithms in these enterprise systems [7]. The trustless-trusted distinction arises again as enterprise systems are often *trusted* rather than *trustless* (such as public Blockchains Bitcoin and Ethereum). Trusted means having to know and trust the involved parties as opposed to only having to trust the software system.

Another reason to distinguish between Blockchains and distributed ledger technologies as the more general form of the technology is that next-generation systems may not use blocks. These systems argue that the requisite level of cryptographic security can be maintained while not having to use a hash-chained blocks architecture like Bitcoin and Ethereum. There may be a trend toward a series of potential shifts in the architecture of transaction execution, to streamline and segment Blockchain networks. The overall effect could be a further instantiation of algorithmic trust (built into network operations) as opposed to physical world trust mechanisms for identifying the parties executing transactions. For example, the notion of a Blockchain (a sequential block of transactions cryptographically hashed together) is being superseded by some projects calling for "blockless blocks," in the sense that a distributed ledger may be cryptographically maintained with hashes that call each other without having to have a block structure. A related method is directed acyclic graphs (DAGs) which do not have sequential blocks of transactions. An example of DAGs is the IOTA project, which uses the architecture of a tangle of transactions. Internet-of-things entities (machine-to-machine transactions) using this network do not need the very high security of financial transactions on the Bitcoin network and have a much smaller peer-based proof-of-work consensus mechanism that consumes much less electricity than the Bitcoin-mining operation. Instead of transactions paying a transaction fee, the network is free and runs via peer-to-peer services. Any node submitting a new transaction for confirmation is asked to conduct a small proof of work to confirm two other random transactions on the network. Another proposed structural mechanism that takes advantage of network properties is path-based settlement, which may provide a more efficient method of transfer with greater privacy [8].

2.2.2 Key Properties of Blockchain Technology: Ledger, Distributed, Immutable

The key properties of Blockchains are how the software operates to conduct the secure transfer of unique instances of value (e.g., money, property, contracts, and identity credentials) via the Internet without requiring a third-party intermediary such as a bank or government. The crucial properties of (base case) Blockchain technology are that it is trustless, distributed (peer-based), consensus-driven, and immutable. First, by *ledger*, what is meant is something like a large online database or Google or Excel spreadsheet, in which rows are added or appended. A row or cell (a "Blockchain cell") can contain any kind of digital information such as account balances (money) or assets, or could also store more sophisticated information such as the memory and execution state of a computer program. All of these data values are stored in the distributed ledger system.

Second, by *distributed*, what is meant is that there is a decentralized network. There are several nodes or peers that are connected in a network such that there is no single point of failure or centralized control. The point is to design a decentralized network so that if some peers crash or attack the network maliciously, the network can still operate. This is the solution to a distributed network computer science problem and is typically called the

Byzantine General's Problem (going back to the Byzantine Empire). Byzantine Fault Tolerance is the ability of a distributed computing system to be fault-tolerant in the case of node failures, malicious, or otherwise. The benefit of a fault-tolerant distributed network is that it creates a trustless system, in which users can trust the technology as a whole, and not human intermediaries. The computing network confers trust, not just by being a software system for executing transactions, but further, in that the technology is mathematically provable as safe and valid in processing these transactions through the cryptographic algorithms. Since transactions do not pass through a central entity, they are censorship resistant. This means that no central authority can prevent any party from conducting business on the network. Third, ledger systems are *immutable* in that once data is committed to the ledger, it cannot be changed. This makes the data auditable to any party who has access to the ledger records. If changes need to be made, they are done so through a second transaction to adjust the first one, but the first one itself cannot be altered (unlike a bank or credit card chargeback).

2.2.3 Cryptographic Identities

The users of Blockchain transaction networks are cryptographic identities, a notion which is conceptually analogous to an email address. For email, you have your address, and to access it you need a password. Similarly, with a cryptographic identity, your public key is like your email address and your private key is like your password. Your software wallet signs transactions, creating a one-time digital signature with your cryptographic identity, which is broadcast to the network and constitutes the account ownership function of being able to execute transactions with your account on the Blockchain. How things function is that it is actually more complicated in that the public key is not sent across the Internet as your address but rather a hash of your public key, which keeps it even safer. Wallets often generate a new address for each transaction, and keep track of them all and which ledger accounts the private keys control. Further, confidential transactions are a new phase of Blockchain technology that is being implemented across platforms including Ethereum, and is already used in Monero and Zcash. Confidential transactions use technology such as ZKPs which mask the sender and receiver address and the transaction balance, cryptographically confirming the balance transfer but not broadcasting the details to the network. The cryptographic identity means that only you can make and sign transactions, which is how your account is kept secure on the public ledger and cannot be accessed and controlled by other parties.

2.3 Consensus Algorithms: POW, POS, PBFT

To update the ledger, the network needs to come to consensus using an algorithm. Arriving at consensus on a distributed network means that everyone agrees on the current state of the ledger (e.g., how much money does each account have) and confirms that no one is double-spending their money. Coming to consensus is a computer science problem in fault-tolerant distributed systems. Generating a consensus means that multiple servers on the distributed network agree on the current truth state of the system, or in the basic Blockchain case, values in the ledger. Once the network computers reach a decision on a value, that decision is final. In the classical computer science context, consensus algorithms are used to agree on the commands in the logs of the distributed servers. In Blockchain networks, the three main kinds of consensus algorithms for arriving at consensus in a distributed manner are Proof of Work (POW), Proof of Stake (POS), and Practical Byzantine Fault Tolerance (PBFT). The main innovation of the Blockchain protocol is the Blockchain data structure on top of a consensus algorithm, which makes it possible to build an open distributed network in which all of the parties can reach agreement.

In POW consensus, the algorithms that operate the distributed system reward miners (client machines on the system) who solve mathematical problems. New transactions executed on the network are routed from the software wallets executing them to all of the mining clients on the network. In Bitcoin, for example, each mining client has a memory pool (*mempool*) to collect incoming unconfirmed transactions. The mining software validates the new transactions and batches them into a block (every 10 min in Bitcoin, and every ~12 s in Ethereum). The financial incentive for each mining client is to be the one to create the new block of transactions that the other machines will all take as the new truth state of the network and append to the decentralized copy of the Blockchain they maintain on their peer node. It is lucrative to record these transactions ("discover a block" or "mine a block"), and thus many dedicated mining operations (running custom ASICs) exist for Bitcoin. Proofs of Work consensus mechanisms are criticized for the wasteful use of computation to produce the cryptosecurity.

In POS, the consensus algorithm is instead based on owning a stake in the network. In POS systems, the creator of a new block is chosen in a deterministic way, depending on its stake, or degree of commitment (wealth) in the network. However, POS systems are perhaps unnecessarily convoluted with voting tiers and not necessarily more efficient than Bitcoin in terms of what might be considered wasteful computation. Therefore, some of the next-generation Blockchains are considering PBFT as a longer-term solution that would allow 1 million distributed machines (clients) on a network to come together in a common and secure truth state of the system. PBFT, which is the ability of a distributed computing system to operate irrespective of faulty nodes (malicious or otherwise). PBFT relies on there being a diversity of participants on the distributed system (a few hundred). For each block of transactions, algorithms randomly select a small, one-time group of users in a safe and fair way. To protect them from attackers, the identities of these users are usually hidden until the block is confirmed. The size of this group typically remains constant as the network grows. For example, 250 of 300 machines (chosen at random per the algorithmic system) would need to sign any transaction whether there are a thousand or a million nodes on the network. DFINITY and Algorand are examples of next-generation Blockchain projects with PBFT consensus algorithms.

3. PUBLIC BLOCKCHAINS: ETHEREUM
3.1 Ethereum Decentralized Application Platform

This section discusses Ethereum, drawing from the Ethereum Project [9], BlockGeeks [10], StackExchange [11], the Berkeley Blockchain Club [12], and Silicon Valley Ethereum [13]. Ethereum is a decentralized platform designed to run smart contracts (computational registration and execution of contractual arrangements). Ethereum is a public Blockchain network for which anyone can download a wallet and use. There is no single point of control or failure, and Ethereum is censorship-resistant. Ethereum is a Blockchain application platform, an overlay to the Internet that is another layer and platform upon which distributed applications can be built.

From a computer science point of view, Ethereum is a distributed state machine in which a block of transactions is equal to a state transition function. A new block of transactions constitutes a state transition function from state A to state B. Thus, Ethereum is like a giant virtual state engine, or decentralized computer, in which multiple parties are sharing the computing platform and running their programs on it. The computing network forms a world-scale decentralized computer. Ethereum has a native asset or cryptocurrency called *ether* which is the basis of value in the Ethereum ecosystem. The purpose of having a cryptotoken is to serve as a basis for transacting, and to align incentives. Ether is also used to provide a reward to miners for finding new blocks (e.g., confirming and recording transactions) and contributing to the security of the network.

3.1.1 Bitcoin and Ethereum Comparison

Whereas Bitcoin is a cryptocurrency (with a basic scripting language), Ethereum is a full-fledged platform. Ethereum is a smart contract execution platform, and Bitcoin is a transaction execution platform. This means that Ethereum is a platform for building distributed applications: applications running on Ethereum as a platform. Bitcoin is a closed destination. It is designed to do one thing, execute one-time monetary transactions securely. Bitcoin is designed to be simple and robust, with high security and high fault tolerance. Complexity is not part of Bitcoin's design goal and therefore Bitcoin's smart contract language is simple and does not afford the same amount of complexity as Ethereum. As an application platform, Ethereum is designed to be complex and feature rich. Another difference is that with Bitcoin, the Bitcoin asset is the center of the Blockchain's activity. In Ethereum, ether is an asset but not the primary goal of the system, but more of an economic mechanism aligning incentives for operating smart contracts.

Ethereum has a Turing-complete scripting language as opposed to Bitcoin's scripting language which is more basic (though is still extensible enough to include multisig functionality (multiple signatures) and prototypical payment channel functionality (with the opcode primitives *nLockTime*, *CheckLockTimeVerify*, and *CheckSequenceVerify*). Conceptually, smart contracts are programs written to be executed on the Ethereum platform. A smart contract instantiates a contract in programmatic code such that the code execution automatically enforces the terms of the contract (as long as it is run by a trusted entity (such as a *trustless* Blockchain network)). The definition is that a smart contract is code that facilitates, verifies, or enforces the negotiation or execution of a digital contract. A trusted entity must run this code, and the Ethereum network executes the smart contracts without any third party needed.

Ethereum has a Turing-complete scripting language which is significantly more powerful than the Bitcoin scripting language and enables smart contracts. Ethereum is Turing-complete in the sense that programs can be written with the full expressive power of any modern programming language. With the Ethereum computing platform, developers can program decentralized applications (DApps) as an alternative to centralized applications. DApps offer a nonproprietary solution with enhanced cybersecurity since there is no centralized database to be a target for hackers. Examples of DApps as the analog to centralized apps include Steemit (decentralized news), Synereo (decentralized social network), LaZooz (decentralized Uber), Ujo Music (decentralized Spotify), and Ethlance (decentralized contractor job listing marketplace).

3.2 Consensus, Transaction Confirmation, and Finality

In Ethereum, the block creation time is every 12s as opposed to Bitcoin where it is 10 min. This means every 12s in Ethereum, there is a new transaction block. The much shorter blocktime in Ethereum means that transaction execution is nearly instantaneous. Ethereum uses what is called the *ghost protocol* to orchestrate consensus and provide the incentive structure to make such a short blocktime possible. The size of public Blockchains (Dec 2017) is that the entire ledger of all transactions since inception is ∼180 GB for Bitcoin and ∼225 GB for Ethereum. However, to run an Ethereum node, only 20–30 GB of space is needed.

Bitcoin and Ethereum are different in terms of the consensus algorithm employed. They both use POW, but different kinds, and Ethereum plans to move to a POS or a hybrid POW/POS consensus mechanism in 2018 (already deployed on the test net in Jan 2018). Where Bitcoin uses SHA 256, the current RSA cryptography standard for POW, Ethereum uses a mechanism called *ethash*. The benefit of ethash is that it is ASIC resistant. Ethash is memory hard (e.g., memory bound), meaning that it is not possible to build an ASIC for it [14]. The implication of not being able to build an ASIC for the mining operation is that Ethereum's consensus algorithm must be run on distributed computers and would not be conducive to the centralized mining farms that orchestrate the Bitcoin mining operation. This means that Ethereum has a stronger claim for being a fully distributed, decentralized, and peer-participation network vs Bitcoin, where miners need to have specialized hardware to participate in the lucrative consensus process. Bitcoin will likely continue to use POW, even though it is ecologically damaging by being computationally expensive (estimated to consume as much electricity as the country of Ireland [15]). Ethereum announced a POS method, Casper, on the test net in January 2018, but it is unclear if the computational cost will be any different than POW.

3.2.1 UTXO (Bitcoin) vs Accounts (Ethereum)

Another difference between Bitcoin and Ethereum is that Ethereum is an account-based Blockchain, and Bitcoin is a UTXO-based Blockchain (Unspent Transaction (tx) Outputs). In Bitcoin, all inputs to a transaction must be in the UTXO database for it to be valid. UTXOs are the unspent amounts from previous transactions that need to be confirmed as unspent to be used as inputs the current transactions. Instead of UTXOs, Ethereum uses an account-based model. In Bitcoin, a user's available balance is the sum of unspent transaction outputs controlled by their private keys. In Ethereum, a user's available balance is recorded in the user's account. The account has the user's cryptographic entity address, balance, and any data in the optional code field. For example, in Bitcoin, Alice owns private keys which control a set of UTXOs. In Ethereum, Carol owns private keys that control an account comprised of an address, balance, and a code field. Ethereum is more efficient than Bitcoin by saving space. By using an account model instead of a UTXO model, Ethereum nodes only need to update their account balance instead of storing every UTXO. Ethereum is also more intuitive in that smart contracts are a more effective programming mechanism to transfer balances between accounts as opposed to constantly updating a UTXO set to compute a user's available balance.

There are two Ethereum account types: externally owned accounts (EOAs) and contracts. All accounts have an address and an ether balance. The first account type is *EOAs*. These accounts are generally owned by some form of external entity such as a person or corporation, the kind that any new user would own when signing up to the Ethereum network. EOAs have a cryptographic address and an ether balance. They can send transactions (transfer ether to other accounts or trigger contract code). The second account type is *contract accounts*, known as contracts. These accounts have an address, ether balance, and any associated contract code. The code execution is triggered by transactions or messages (function calls) received from other contracts or EOAs. Contracts are programs that are smart contracts (autonomously operating code) and have persistent storage. Code execution is triggered by transactions or messages (function calls) received from other contracts or EOAs. This means that contracts are autonomous accounts on the Ethereum network that other accounts (contracts and EOAs) can interact with them, but no one controls them, because once launched, they are autonomous. It may be possible to interact or execute some transactions with a contract since other programs can call functions on it, but otherwise a contract account cannot be directly controlled. The way that accounts interact with the network, other accounts, other contracts, and the contract state is through transactions.

The accounts on the Ethereum network comprise the network state. This means that the state of *all of the accounts* at the same time together is the state of the Ethereum network. The entire Ethereum network agrees on the current balance, storage state, and contract code of every single account, updating the state machine with each block (every 12 s), which is why consensus is expensive. Each new block updates the overall state of the Ethereum network, taking the previous block and creating a new block. The new block takes the previous state and produces a new network state. Each network node must agree on the new network state. Thus, the blocks are the state transition function between the states.

3.3 The Ethereum Virtual Machine

The central element of the Ethereum platform is the Ethereum virtual machine (EVM), which serves as the execution model for smart contracts. The EVM runs contract code in the sense that the contract code that is executed on every node is EVM code. The contract code is not written in the high-level Turing-complete programming language, but rather in a low-level simple stack-based programming language that looks a lot like JVM's bytecode (the java virtual machine). Every Ethereum node runs the EVM as part of its block verification procedure. This means that for participants in the Ethereum network, each node is an independent instance running the EVM code as the mechanism for how each node verifies that new blocks are valid, and that the computation has been done correctly. Since every node is doing all of this computation, this might not be the most efficient final model, but like Bitcoin, it has high cryptographic security.

More technically how the EVM operates is as a state transition function. How it works is that a string of parameters is fed into the EVM to generate the new *block_state* of the overall network and the new amount of *gas* that accounts have. The input string has several parameters as it goes into the EVM and limited outputs: (block_state, gas, memory, transaction, message, code, stack, pc) $\rightarrow$ EVM $\rightarrow$ (block_state, gas). The *block_state* is the current state of the Ethereum network, the global state containing all of the accounts including balances and long-term storage. The *gas* is the cost of computation which are the fees required to run these computations (they vary by the amount and type of computation). The *memory* is the execution memory. The *transaction* is the transaction itself. The *message* is the metadata about the transaction. The *code* is the code itself. The other parameters are execution related, the *stack*, and the *pc* (program counter). This sequence of opening parameters is fed into the EVM to generate the new *block_state* of the overall network and the new amount of *gas* that accounts have.

The design goals of the EVM are fivefold: simplicity, space efficiency, determinism, specialization, and security. The EVM is designed to be simple so that it can easily prove the security of smart contracts, which also helps to secure the platform itself. Considering the full stack, the idea is that the smart contracts will be complex and feature rich, so keeping the virtual machine simple allows the complexity to be layered on farther up the stack. The EVM is designed to be space efficient meaning that the EVM assembly is as compact as possible. The EVM is deterministic in the sense that the same input state should always produce the same output state. Having a deterministic virtual machine necessarily restricts the scope of activities, for example, an HTTP request of Ethereum is not possible. However, the EVM is designed to be specialized so that it has built-in functionality for expected operations such as cryptographic functions (easily handling 20-byte addresses and custom cryptography with 32-byte values), modular and exponential arithmetic (used in custom cryptography), and to read block data, transact block data, and interact with the *block_state*. The security of the EVM is designed so that it is easy to create a gas cost model for operations so that the virtual machine is nonexploitable. The main programming language for the EVM is Solidity (Serpent is another). The process is to program a smart contract in Solidity and then send it through the Solidity Compiler (solc) to generate the EVM code (likewise writing a Serpent Contract and sending it through the Serpent Compiler to generate the EVM code). The complexity of the contract language is managed through the Solidity Compiler and the Serpent Compiler, but at the architectural level, remains a simple stack-based language.

3.4 Ethereum's Gas: The Cost of Computation

To prevent Denial of Service Attacks and address the halting problem (in which the network gets stuck trying to execute a contract that produces an infinite loop), Ethereum's solution is *gas*: every contract requires *gas* to fuel the contract execution. Every opcode in the software requires some amount of gas to execute and the more expensive a certain opcode is, the more gas it needs to run. For example, a simple opcode to *add* a record will require very little gas, whereas cryptographic exponentiation might be significantly more expensive (in fact, thousands of times more expensive). Each transaction will specify the *startgas*, which is the maximum amount of gas that you are willing to spend on this computation, and the *gasprice*, the fee per gas that you are willing to pay. Even in the slightest network operation, economic principles are involved in terms of the cost of computational resources, and the users demand to use these resources.

At the start of the transaction, *startgas* multiplied by *gasprice* is computed, and that is the amount of ether that is subtracted from your account. If the contract executes successfully, then the remaining gas that was not taken from *startgas* is refunded back to you, but if the contract runs out of gas during execution or otherwise fails, then the execution reverts, and you do get refunded all of your gas. Therefore, you should always specify more *startgas* than you think you will need. Technically, Ethereum still allows an infinite loop, so a DoS attack is possible, it is just prohibitively expensive. An attacker can write the code, but presumably they do not have infinite resources to pay for an infinite loop to produce the DoS attack. The idea is that buying gas is like buying distributed trustless computational power.

Overall, the goal of Ethereum is not to optimize the efficiency of computation. Ethereum is a massive parallel processing system. It is specifically redundantly parallel because every node verifies each block, meaning that every node needs to run all of this code. In this manner, the Ethereum network achieves consensus on the state in a distributed network without needing a trusted third-party intermediary. The Ethereum system is truly distributed (with all of the benefits of Blockchains (trustless, censorship-resistant, immutable)), but it means that each node has to run all of the code that produces state changes to the overall system. Running code on Ethereum is expensive because there is one global network and everyone is putting their smart contracts onto this network, and has to share the same resources, because in the end there are always bottlenecks of the computational resources of any one computer. A mixed strategy would put the features of the Blockchain to the best use. The idea is to perform computation off-chain and then secure it on-chain (record important values with a time-date stamped hash). Business analytics, machine learning, and data storage would not be executed on the Blockchain, but salient values obtained from these processes might, particularly for interaction with smart contracts.

4. PRIVATE BLOCKCHAINS: CORDA, ETHEREUM QUORUM, HYPERLEDGER, RIPPLE

4.1 Enterprise Blockchain Systems in Financial Services

This section provides an overview of private Blockchain systems intended for use in business, industrial, and government use cases, particularly a look at the technical specifications, operation, and implementation of enterprise Blockchain solutions in the first and largest sector of their implementation, the financial services industry. The enterprise-class financial services industry private Blockchain projects are in the process of being realized by groups of financial institutions and enterprises with R3's Corda, the Linux-led Hyperledger consortium, and the Enterprise Ethereum Alliance (deploying Ethereum Quorum). Seeing the implementation in the financial services sector demonstrates how some of the same features and structures may transfer to implementation in other sectors. The examples discussed here draw upon proof-of-concept projects undertaken by Accenture and the Monetary Authority of Singapore [16], R3 and Bank of Canada's Project Jasper [17], JPMorgan Chase Quorum [18], and Credit Suisse, Finastra, HQLAX, and R3 [19]. As a caveat, Blockchain systems are at a stage of rapid change and any technical description included here may be outdated as solutions evolve, implementations become more robust, and known issues resolved.

In the private enterprise Blockchains, the biggest activity so far is in financial services, where distributed ledgers are a next generation in the ongoing progression of Fintech (making financial operations more effective with technology). The biggest implementations in the financial services industry so far are with R3's Corda, Symbiont's smart ledger platform, Ethereum's Quorum, and Ripple. Quorum is based on Ethereum, Symbiont is based on Counterparty, and Corda and Ripple are home-grown. After the financial services industry, the next big groups of adopters are industry, manufacturing, and supply chain, and energy. Another important sector for Blockchain implementation is healthcare and clinical trials where the killer apps are electronic medical records (EMRs), electronic consent managed by smart contracts, and harmonized, interoperable, standardized data, and using the privacy–transparency properties for health research.

Enterprise Blockchain systems typically have some but not all of the core properties of public Blockchain systems. The core properties are that public Blockchains are distributed, trustless, and consensus driven. Enterprise Blockchains may be distributed in the sense of having a decentralized network architecture with multiple different kinds of peer nodes, but they may be quite different from public Blockchains regarding the parameters of being trustless and consensus driven. Most enterprise Blockchains are not trustless, they rely on having to know and trust the parties controlling the Blockchain and issuing credentials and privileges. Especially since most of the activity on private Blockchains may be hidden, it is incumbent on Blockchain users, counterparties, and regulators to understand precisely how transactions, private views, and changes to the underlying system operate.

Transaction settlement confirmation and finality is a crucial aspect of enterprise Blockchains, especially in the financial services industry in processing securities. It is important to understand the variety of settlement confirmation and finality processes in the different Blockchain systems. For example, in Bitcoin, transaction blocks occur every 10 min, but for finality, it is recommended to wait up to 1 h, since each 10-min block of time that goes by is added to more peers in the ~12,000-node peer network (Mar 2018), geometrically expanding the assurity of the transaction settlement and finality. It is not like a commercial bank, where being centralized in the classical model, there is surety in near-immediate finality. In securities transactions, settlement finality is a specific statutory, regulatory, and contractual construct [20]. In general, settlement finality refers to the moment in time when one party is deemed to have discharged an obligation or to have transferred an asset or financial instrument to another party. At this moment, the discharge or transfer becomes unconditional and irrevocable.

4.2 Corda From R3

Corda is a distributed ledger platform for use in the financial services industry made by R3, a distributed database technology company. On the Corda platform, the ledger is distributed on a need-to-know basis, as opposed to a global broadcast method. This means that only the parties involved in transactions have visibility into these transaction details. Confidential identities is an additional layer of privacy in the Corda system such that only transaction counterparties can identify the participants. How this works is that transaction participants exchange a new key and certificate using Corda's "Swap Identities Flow." Newly issued keys are used for the detailed specification and signing of the transaction. Although the sender and receiver details are anonymized with confidential identities, the amount of the transaction is not. This is because the transaction amount enables the Corda traffic management algorithm to compute the best routing of the transaction based on the amount of the queued payment instructions (which could result in a lack of net neutrality principles). The risk of exposing the transaction amount is that it might be possible for network members to graph the network and deduce the sender and receiver of each of the queued payment instructions and the aggregate amounts of transfer.

This is important because research is revealing cryptocurrency transactions to be less private than might have been thought given the domain's pseudonymous wallet addressing system. For example, Blockchain analytics firm Chain analysis claims that since information related to 25% of Bitcoin addresses is tied to real-world identities, it can account for approximately 50% of all Bitcoin activity [21]. Teams have back-calculated or identified 50% of Bitcoin transactions, 78.7% of Ripple transactions [22], and 62% of transaction inputs in Monero transactions [23].

Corda is easy to use and plug-and-play with existing corporate information technology architectures. Adding a new node in Corda involves only the installation of the new node itself, and minimal change to existing nodes and the overall network. Since Corda transactions are only sent to nodes on a need-to-know basis, each peer only sees their relevant subset of total ledger transactions. This makes Corda performance more efficient than classical distributed Blockchain systems since not every peer node is confirming every transaction, and storing and updating the entire ledger. However, it also means that Corda is more of an enterprise software system than a truly distributed Blockchain system. The benefit of Corda as an enterprise system with some attributes of Blockchain technology is that some of the scalability and performance issues commonly experienced in classical distributed ledger systems may be eliminated.

The "consensus mechanism," or more specifically the transaction validation, confirmation, and logging mechanism Corda uses is a Simple Notary service. If only one Simple Notary service is used, a single point of failure is created in the network. If the single Simple Notary service fails, transactions cannot be completed. To reduce risk, the Simple Notary service might be implemented as a cluster that is potentially operated by multiple parties. Such a solution can be designed to support load balancing to increase transaction throughput, multithreading of incoming transactions and minimizing latency for geographically diverse transacting parties. In industries where high transaction throughput is a requirement, multiple notary services might decrease risk and allow greater transaction throughput and load balancing.

Corda (like Bitcoin) uses a UTXO model that requires input states to be linked to one or more inbound transactions by their hash, in order to produce an immutable chain of asset lineage. Calculating the asset lineage chain in every transaction can become a performance issue as each involved node in any transaction verifies that every input was generated in a sequence of valid transactions to confirm the overall authenticity of the chain. Performance scalability might be enhanced with best practices that periodically expire and reinstantiate digital assets, or recycle unused pledged funds after a period of time. If any of the participating nodes are unreachable, for example, due to machine failure, network issues, or other causes, the network can still operate for all transactions that do not require the involvement of the failed nodes. At the same time, nodes can be shut down and restarted at will due to Corda's *Flow Checkpointing* mechanism, ensuring data is never lost and that flow progression is protected. In addition, fund transfer transactions can still proceed among participating nodes that are not involved in off-line nodes. Valid fund transfers can be initiated and settled in real time.

In Corda, the notary service provides the point of finality in the transaction. The presence of a notary signature indicates the transaction's finality (so finality occurs at the moment of notary transaction signing for ownership and liability purposes). By obtaining a notary signature, transaction participants can be sure that the input states are unconsumed (unspent) by prior transactions. The notary model is structured for atomicity (a system property of not allowing transaction splitting into pieces (atoms)). In Corda, atomicity means that a proposed transaction is accepted on an all-or-none basis (so there are no partial transactions). Prior to execution, queued instructions for transaction processing can be modified (reprioritized or canceled). Modification is the mechanism for planned transaction failure. To stop unexecuted transactions, the atomicity property of the Corda netting solution can be used so that transactions will fail either because the graph is no longer valid due to modified payment instructions or per a decrease in balance.

4.3 Ethereum Quorum

Quorum is an enterprise-focused version of Ethereum. Quorum is a private Blockchain enterprise-ready distributed ledger and smart contract platform. Quorum is good for any application requiring high speed and high-throughput for the processing of private transactions within a permissioned group of known participants (for example, a group of investment banks). More specifically, Quorum supports both public and private transactions within a permissioned network. The public transactions are broadcast to all the nodes within the network and are processed like regular Ethereum transactions. The private transactions are sent directly to the specified recipients by Quorum's privacy service Constellation as encrypted blobs. It does this by sending the transaction payload only to the involved participants (the rest of the network sees only a hash of the encrypted payload). The reason for propagating transaction hashes to all participants is to easily confirm any transaction in the future. The Blockchain architecture means that if any party to a private transaction needs to validate of the existence of that transaction in the future, they can easily confirm it with the rest of the network by comparing it to the hashes in the network record, thereby not needing to seek or trust information held with the transaction counterparty. In Ethereum Quorum, a smart contract DApp that orchestrates payment flow, manages transaction proof generation and submission, and acts as a link between the public and private contracts.

Quorum allows confidential transactions with ZKPs, which manage the shielded salted balance of each network participant. Balance-masking means that the true balance position of any participant bank is only visible to itself. Any change in balance movement as part of transaction execution can only happen via the submission of a ZKP by the sender and receiver, followed by verification of these proofs by the entire network. This allows transaction and balance validation by the decentralized network without knowledge of the true balance and thus avoids double spending. *Salting* refers to a network defense strategy to protect against cyber-attackers engaging in a *Rainbow table* attack (using a precomputed table to attempt to reverse cryptographic hash functions, for cracking password hashes) [24]. A *salt* is a randomly generated set of additional characters inserted along with a password or address hash to make it harder for cyber-attackers or network participants to guess transaction data. Ethereum Quorum uses dynamic salt to hash transaction amounts, starting balance, and ending balance, uniquely for each transaction. Additional protection is afforded in that payment instructions can only be executed with the specified receiving party. Payment transfers on the public contract identify the sender and receiver but shield transfer amounts by storing hash values. Account balances are kept in private contracts and are only accessible from the account owner's node.

Ethereum Quorum uses the Raft consensus model, which is equivalent to Paxos, a known fault-tolerant high-performance consensus model for distributed computing systems. With Raft, nodes are either *leaders* or *followers*. There is a single leader for a cluster of operations, which all log entries must flow through. At present leaders are elected and persistent. Future Quorum releases may include a Byzantine fault-tolerant consensus mechanism such that the leader is rotated before every block creation. In the public Blockchain implementation of Ethereum, there are not leaders and followers. Any node in the network can mine a new block and become the "leader" for that round. Whereas in the public Ethereum Blockchain, the node roles are minter (recorder of a new block) and verifier (the other peers that verify the block), and in Ethereum Quorum, the roles are leader and follower (new block minter and new block updaters). The articulated reason to have a leader in Ethereum Quorum is architectural, to avoid having a network hop from a node minting blocks to the leader, since the network is set up such that all Raft write entries must flow through the leader. The existing Ethereum P2P transport layer is used to communicate transactions between nodes, but blocks are only communicated through the Raft transport layer. Blocks are created by the minter and flow to the rest of the cluster, always in the same order, via Raft. When the minter creates a block, unlike in the public Ethereum Blockchain in which the block is written to the database and immediately considered the new head of the chain, in Raft, the block is only inserted or set to be the new head of the chain once it has flown through Raft. All nodes extend the chain together in lockstep by applying the Raft log.

In implementing Ethereum Quorum, the Raft-based consensus can be set up in a one-to-one correspondence between Raft and Ethereum nodes such that each Ethereum node is also a Raft node. In this architecture, the leader of the Raft cluster is the only Ethereum node that can mint new blocks. A minter is responsible for bundling transactions into a block just like an Ethereum miner, but does not present a POW. The Raft leader commits new blocks to the chain after verifying the block's transactions and all followers update to the latest block in lockstep. Once a block is committed to the chain it cannot be reversed, which provides transaction finality. The Raft leader is elected during network creation. When the nodes in the network detect that the Raft leader is down, the network elects a new Raft leader, allowing transactions to continue to be processed. Quorum uses the same block propagation mechanism as Ethereum. This means that if any node goes down or is disconnected from the network, the rest of the network still functions as usual, but transactions with the disrupted node will not be completed. When the disconnected node comes back online, the transaction history is automatically synchronized.

4.4 Hyperledger Fabric

Hyperledger Fabric is a Blockchain framework implementation hosted by the Linux Foundation. The Hyperledger Fabric framework is intended to be a foundation for developing applications with a modular architecture, by allowing components such as consensus, membership services, and business analytics to be plug-and-play additions to existing corporate IT infrastructure. Hyperledger Fabric uses standard container technology to host smart contracts written with *chaincode* that comprise the business and application logic of the system. Hyperledger Fabric was initially contributed by Digital Asset Group and IBM.

Hyperledger Fabric is a permissioned network with the ability to set up private channels between participants in which each channel maintains an independent ledger. Channels allow information to be shared between parties on a need-to-know basis. A channel is a data-partitioning mechanism which limits transaction visibility to participants. Other network members cannot access the channel and are not able to see transactions in the channel. This enables the setup of ledgers which (1) can be shared across a large network of peers (called a netting channel) or (2) can be defined to only include a specific set of participants (bilateral channels). By design and to emphasize privacy, since transactions and queuing mechanisms occur within bilateral channels, transaction details are only visible to the involved pair of parties in the bilateral channel. Both parties in the channel can view the transactions of both party's channel-level accounts. However, since any party can only view the balance of one channel-level account per counterparty, it is not possible for any party to deduce the total balance or liquidity of a counterparty within the ledger channel.

The transaction channel structure of Hyperledger is intended to facilitate the movement of funds and transactions across channels. Depending how the system is configured, it may be possible for network participants to identify the channels involved in each transaction. Thus, privacy may be improved by securing data at the infrastructure level, such as by deploying additional cryptographic functions to ensure confidentiality. It might also be possible to remove the funding channel altogether to allow for higher efficiency. The netting channel has information regarding the identities of the transaction participants, transaction execution instructions, and the net value of nettable payment instructions for each party. Individual transaction amounts are not exposed. In a transaction cycle with limited transaction instructions, it may be possible to deduce the amount of the transaction. In other higher flow cases in which netting is likely to involve multiple transactions per cycle, the amount of individual transactions might be less deducible. One solution to this privacy issue is bilateral channels. Bilateral channels allow bilateral netting to happen within a channel, without involvement of the rest of the network participants. This means that overall network traffic flow may not be a bottleneck as introduces the possibility that queued payment instructions in a bilateral traffic can be resolved within the bilateral channel without having to depend fully on traffic resolution with multiple parties in the network. The design of an individual bilateral channel between pairs of transacting banks also introduces the need for operators to maintain funds or other transaction resources in each channel. Transaction execution that conforms to business operation rules might be automated with chaincode smart contracts.

For the purpose of a proof-of-concept project, a Hyperledger Fabric setup could consist of one orderer which sends transactions to all peers in a channel for validation. A multiple node ordering service (such as Kafka) can be implemented for high availability of the ordering service to ensure it does not become a single point of failure. Kafka is an open-source stream processing platform that allows high-throughput and low-latency processing for real-time data feeds. The overhead caused by the setup of multiple channels in this design can be reduced by means of sharing hashed data among all participants while keeping private data with a limited set. This new component is underway and expected in future releases of Hyperledger Fabric. When transitioning from proof-of-concept to production phases of Blockchain implementation projects, there would likely be more than one ordering service. This is because having only one ordering service that sends transactions to all peers in a channel for validation would be a single point of failure for the network. If the orderer fails, there could be disruption as transactions will not be ordered into blocks and committed to the chain. This issue can be resolved with the implementation of multiple node ordering service (e.g., Kafka) for high availability. One trade-off in the Hyperledger Fabric workstream design is the number of channels required to guarantee high levels of privacy, while achieving traffic resolution. Thus, cross-channel communication (e.g., movement of funds from one bilateral channel to another) becomes part of the design. The coordination logic between two channels needs to be programmed in a custom application using a framework such as Node.js since cross-chain interaction is not yet supported. In early development phases, it could be good to have a rollback mechanism for an application-orchestrated function that involves multiple chaincode executions so that its operation may be traced. Also in prototyping phases, it could be good to have redundant nodes setup in the systems architecture to support higher resiliency.

Hyperledger uses a broader conceptualization of consensus than the classical Blockchain definition of consensus in public Blockchains as a specific algorithm with a single function of cryptographic validation in transaction confirmation. In Hyperledger Fabric, consensus is used in the full transaction flow, from proposal and endorsement, to ordering, validation and commitment. In Hyperledger Fabric, consensus is the full-circle verification of the correctness of a set of transactions comprising a block. Consensus is achieved when the order and results of a block's transactions have met the explicit policy criteria checks. These checks and balances take place during the lifecycle of a transaction and include the usage of endorsement policies to dictate which specific members must endorse a certain transaction class, as well as system chaincodes to ensure that these policies are enforced and upheld.

Prior to commitment, peers use system chaincodes to ensure that enough endorsements are present and are obtained from the appropriate entities. (There are three node roles in Hyperledger Fabric, orderer, endorser, and committer.) In addition, a versioning check takes place during which the current state of the ledger is agreed or consented upon, before any blocks containing transactions are appended to the ledger. This final check provides protection against double-spend operations and other threats that might compromise data integrity, and allows for functions to be executed against nonstatic variables. In addition to the endorsement, validity, and versioning checks that take place, there are also ongoing identity verifications happening in each direction of the transaction flow. Access control lists are implemented on hierarchical layers of the network (ordering service down to channels), and payloads are repeatedly signed, verified, and authenticated as a transaction proposal passes through the different architectural components. For users of Hyperledger business Blockchain networks, it is important to check how these consensus processes are happening and who can change them and how. Regarding transaction settlement and finality, the peers defined in the endorsement policy of the chaincode endorse the transaction for each transaction. The endorsed transaction is sent to the orderer, it is ordered into a block, and the block is broadcast to the channel participants to validate and commit to their ledgers. This mechanism assures the finality of transactions within a channel. However, at present there is no chaincode method for transactions with communication between channels, so conventional technologies such as Node.js must be used.

4.5 Ripple

Ripple is a real-time gross settlement system, currency exchange, and remittance network initially released in 2012. It handles both immediate cash payments and credit transactions over time. As of January 2018, it was the fourth largest cryptocurrency, with a market capitalization of $51 billion. Ripple is built on a distributed open source Internet protocol, consensus ledger, and native cryptocurrency called XRP (ripples). The shared public database (ledger) uses a consensus process with validator nodes (credentialed external parties such as MIT) confirming the payment, exchange, and remittance transactions. Transactions may be denominated in traditional fiat currencies, cryptocurrencies, and user-defined currencies such as frequent flier miles or mobile minutes. Ripple's goal is to enable secure, instantaneous, very low-cost global financial transactions of any size with no chargebacks. Ripple is different from other cryptocurrency projects such as Bitcoin, Ethereum, and Monero in that it targets financial institutions, and might possibly be a successor to SWIFT, the current global payment system used among financial institutions.

As of October 2017, over 100 worldwide financial institutions were participating in the Ripple network. Members are 12 of the world's top 50 banks, including Santander, BBVA, and Standard Chartered [25]. Ripple has been received favorably by an initiative of the US Fed, the Faster Payments Task Force, which calls for next-generation global payment systems. A McKinsey study reviewing projects for the Fed task force cites the Ripple network's ability to allow financial institutions to operate cross-border payments much faster than the 2–4 days that is common today, while providing end-to-end transaction visibility and settlement confirmation to banks [26]. The capability for banks to perform cross-currency transactions in a matter of seconds for a small fee in a publicly verifiable manner could substantially improve financial institution cost and risk profiles. Doing business internationally within the current global payments infrastructure often requires financial institutions and corporations to prefund local currency accounts around the world to quickly send payments in a given market. Ripple estimates that $5 trillion in capital is unproductively committed this way and might be freed for other business uses [27].

Ripple is distinct as a cryptocurrency both technically and conceptually. Technically, propagation through the credit network is orchestrated by path-based settlement. This means that transactions literally flow through the network in a dynamic crediting–debiting path between available network computer nodes from sender to recipient. This is different from Bitcoin and other similar cryptocurrencies, in which submitted transactions are sent to the memory pool on each mining client machine and are then packaged into transaction blocks that become part of the permanent ledger. Conceptually, Ripple is both the standard notion of cryptocurrencies as a real-time payment system, *and* more fundamentally, a new class of financial network. Ripple is a credit network in that ongoing trust relationships are maintained between parties on the network which have a specific financial value attached to them. Ripple is a mesh network of open IOU credits. The name Ripple refers to *rippling*, the idea that transactions ripple, or flow automatically across open nodes in the network to their destination. A central premise of Ripple is that transactions can automatically ripple across the network, crediting and debiting intermediary wallet nodes without requiring human intervention. As such, Ripple is a credit graph and stores live credit availability in the edges that connect the network nodes. Open IOU credit balances populate the network to enable future transactions. Credit remains resident in the live credit network, which presupposes trust in its future availability and redemption.

The Ripple network structure is comprised of gateways, market makers, and users. A *gateway* is a well-known business wallet (such as Standard Chartered Bank) that can authenticate and bootstrap credit links to new wallets wishing to join the network. Gateways are the Ripple counter parts of commercial banks and loan agencies in the traditional credit model. Gateway wallets maintain high network connectivity. A newly created Ripple wallet that does not have any trust relationships with other wallets can create a credit link to a gateway, and through this relationship, interact with the rest of the network before forming direct links to other wallets. Gateways may likely enforce a known-identity credentialing process for new wallets for regulatory purposes (e.g., KYC–AML compliance) and to mitigate risk. Ripple wallet identities (though possibly known to their initial gateway) are pseudonymous to the overall network. The identification process before anew credit link is created and funded may reduce the number of credit links in the Ripple network, and gives rise to a regional geographical structure to the network. This results in a slow-mixing, unclustered, and disassortative network. The slow-mixing property is similar to other networks where link creation requires establishing trust, and possibly physical interaction [28]. A *market maker* is a wallet that conducts currency exchange services.

There are two kinds of Ripple transactions: direct XRP payments between parties and path-based settlement transactions. Path-based settlement transactions transfer any kind of credit (fiat currencies, cryptocurrencies, or user-defined currencies) between two wallets having a valid credit path between them. A valid credit path is a network path through wallets that have preestablished lines of credit extended from one wallet to another. A credit path allows transactions to ripple across the network (e.g., flow through a progression of nodes). Edges are undirected in the sense that a dynamic path can be found through network nodes between sender and receiver at the time of the transaction. During the course of the transaction, the credit value on each edge of the path from one wallet to another is updated directionally. Edges in the direction from the sender to the receiver are increased by the amount of IOU credit being transferred, and reverse edges are decreased by the amount, almost simultaneously as the transaction propagates. Edge weights (i.e., the amount of credit availability) must accommodate the amount of the transaction being transferred. A settlement transaction can be split (packetized) into multiple paths as long as credit is available. The existence of a valid credit path between a pair of wallets is enough to execute a settlement transaction between them. Thus, settlement transactions are possible between arbitrary pairs of wallets, even if they have not extended direct credit links to each other.

In Ripple, the consensus process operates in that network nodes share information about candidate transactions. Through the consensus process, validating nodes agree on a specific subset of the candidate transactions to be considered for the next ledger. Consensus is an iterative process in which nodes relay proposals, or sets of candidate transactions. Nodes communicate and update proposals until a supermajority of peers agree on the same set of candidate transactions. During consensus, each node evaluates proposals from a specific set of peers, called chosen *validators*. Validators represent a subset of the network which, when taken collectively, is trusted not to collude in an attempt to defraud the node evaluating the proposals. This definition of trust does not require that each individual chosen validator is trusted. Rather, validators are chosen based on the expectation they will not collude in a coordinated effort to falsify data relayed to the network. Candidate transactions which fail to be included in the agreed-upon proposal remain candidate transactions. They may be considered again in the next round of consensus. Typically, a transaction which does not pass one round of consensus succeeds in the following round. However, in some circumstances, a transaction could fail to pass consensus indefinitely. One such circumstance is if the network increases the base fee to a value higher than the transaction provides. The transaction could potentially succeed if the fees are lowered at some point in the future. The *LastLedgerSequence* transaction field is a mechanism to expire such a transaction if it does not execute in a reasonable time frame. Applications should include a *LastLedgerSequence* parameter with each transaction. This ensures a transaction either succeeds or fails on or before the specified ledger sequence number, thus limiting the amount of time an application must wait before obtaining a definitive transaction result.

As of August 2017, on the Ripple network, there were a total of 181,233 wallets, with 352,420 credit links between them (i.e., far from being completely connected with direct links between each of them) and 29,428,355 total transactions (roughly half XRP direct payments and half credit settlement transactions). At-risk situations might include network liquidity, resilience to faulty wallets, and the effect of unexpected balance changes was investigated. Two risks were identified. First, due to the setting of a software wallet parameter, 11,000 wallets, with a total value of USD $13 million (as of August 2017), could potentially be at risk of being redistributed to lower credit quality IOUs. The issue is that credit rippling through nodes may not be innocuous since lower credit quality IOUs might be substituted in the process. If wallet parameters are not specified in a certain way, as transactions ripple through nodes, a higher-credit quality IOU might be replaced by a lower-credit quality IOU. Wallet nodes are exposed to credit risk in that the credit quality of the IOU could change over time. This risk can be avoided by setting software parameters a certain way, particularly the *no_ripple* and the *defaultRipple* flags. This points out the need for user knowledge and best practices regarding the new domain of digital financial networks. The second risk the study found is that although the overall Ripple network has high liquidity, newly emerging regions might be subject to disruption since transactions flow through local gateways. The study found that one emerging geographical user base of about 50,000 wallets was prone to disruption by as few as 10 highly connected wallets in the region. Again, the issue can be remedied by user awareness and the application of a higher degree of control in wallet software settings.

5. BUSINESS ANALYTICS, ARTIFICIAL INTELLIGENCE, AND DEEP LEARNING CHAINS

Considering the topic of big data analytics for Blockchain, Blockchain both produces and facilitates data standardization and analytics. Blockchain is the disruptive enterprise software of the current moment and the near-future (with 57% of large worldwide firms investigating and implementing the technology [2]). Blockchain adoption is in the bigger frame of contemporary reality that is constituted by big data. Since the beginnings of the Internet, modernity has been marked as regime of big data, with global data volume estimated to exceed 40 exabytes in 2020, equally comprised of scientific, governmental, corporate, and personal data [29]. However, big data is not smart data, the problem is that older algorithms cannot keep up with the growth in data [30]. Thus, new data science methods are needed, those supplied by next-generation artificial intelligence solutions in the form of deep learning algorithms, together with Blockchain-based cryptosecure data which is rendered trustable and interoperable through standardized formats and validation. Blockchain is not just helpful for big data analytics, but necessary. Already convergence of the two emerging technologies are joining in the concept of deep learning Blockchains.

5.1 Deep Learning Chains

The first argument for considering Blockchain and deep learning together suggests the emergence of a new class of global network computing system. These systems are self-operating computation graphs that make probabilistic guesses about reality states of the world. The second reason to consider the convergence of Blockchain and deep learning is that each is necessary to facilitate the development of the other. This includes using deep learning algorithms for setting fees and detecting fraudulent activity, and using Blockchains for secure registry, tracking, and remuneration of deep learning nets as they go onto the open Internet (in autonomous driving applications, for example). Blockchain peer-to-peer nodes might provide deep learning services as they already provide transaction hosting and confirmation, news hosting, and banking (payment, credit flow-through) services. Further, there are similar functional emergences within the systems, for example, LSTM (long–short-term memory in RNNs) is like payment channels. The third reason to notice the convergence of Blockchain and deep learning is the smart network thesis. We are starting to run more complicated operations through our networks: information (past), money (present), and brains (future). There are two fundamental eras of network computing: simple networks for the transfer of information (all computing to date from mainframe to mobile) and now smart networks for the transfer of value and intelligence. Blockchain and deep learning are built directly into smart networks so that they may automatically confirm authenticity and transfer value (Blockchain) and predictively identify individual items and patterns.

5.2 Smart Networks: The Future of Deep Learning Blockchains

The next phase of deep learning could be putting these systems online on the open Internet. So far deep learning has not been a "general capability" and has been confined to hardware clusters tasked on specific projects. However, understanding that *the core functionality of deep learning is being able to identify what something is*, and *self-optimize both its parameters and architecture*, suggests that these could become standard features of any computational system, even the lightest front-ends (browsers, Smartphone, cameras, and IoT sensors). What we have witnessed so far in the world might be understood as the first phase of the Internet, characterized by the transfer of information via simple networks. Today, a second phase of network computing is emerging as we push more complexity and functionality through the Internet pipes, in what might be termed *smart networks*.

Smart networks are computing networks with intelligence built in such that identification and transfer is performed by the network itself through protocols that automatically identify what things are (deep learning technology), and validate, confirm, and route transactions (Blockchain technology) within the network. Blockchain technology is the cryptographic distributed ledger software that enables the secure, end-to-end, and computationally validated transfer of value (whether it is represented by money, assets, or contractual arrangements) via smart networks, as discussed by Swan [3]. Network intelligence is "baked in" to smart networks, and includes deep learning algorithms for predictive identification, and Blockchains to confirm authenticity and transfer value. The benefit of combining deep learning and Blockchain technology is that Blockchain offers the secure audit ledger of activity so that deep learning can be more safely implemented on the open Internet. Overall, the reason Blockchain deep learning systems running on smart networks could be important is that they might deliver an advanced computational infrastructure with which to tackle the next tier of large-scale real-time predictive data science problems such as genomic disease, protein modeling, energy storage, global financial risk assessment, real-time voting, and astronomical data analysis.

6. USE CASES BY INDUSTRY

6.1 Blockchain Business Networks

This section describes Blockchain business networks, and how their implementation might unfold in the corporate, industrial, and government environment. The first industry-wide process change that is a precondition for Blockchain business networks is the digitization of assets. The new mode of business practice may involve registering assets to Blockchains for administration, ownership confirmation, transfer (buying and selling), audit tracking, and compliance. Some of the first kinds of this activity are underway in the form of Blockchain land title registry projects [31,32]. Digitized assets could mean the ability to have a consolidated view of organization assets at various permissioned levels of detail. This could be at the department level or firm level internally within organizations; at the industry-wide level for regulators; and at the economy-wide level for central bankers, risk managers, and policy makers.

Having an asset registered to a Blockchain means that the private key that controls the asset must be used for any transaction involving the asset. Any attempted transactions without the private key would be invalid. Therefore, there is more scrutiny in the audit trail, and it is less likely for the asset to be pledged in questionable or illegal activities. Further, in corporate environments, multiple signatures (multisig functionality) are usually required for asset transactions above certain thresholds, which brings more inspection and audit tracking to the domain of asset management. These requirements suggest that Blockchains may improve the trust and security of business activity. Trading partners would not agree to non–Blockchain registered transactions because they would lack enforcement. Any contractual arrangement involving an asset would have to be registered to a Blockchain otherwise the arrangement could not be enforced.

6.2 Real-Time Asset Valuation

If assets are digitized and registered as Blockchain assets, they might be valued with greater ease, and even automatically. Currently, organization balance sheets are a snapshot of values at a past moment in time and do not reflect present market value. The digital instantiation of assets means that smart contracts or other tools could constantly or periodically value these assets in real time. It has long been possible to see continuous market values for assets such as securities, and now cryptocurrencies. However, for corporate balance sheets, there has not been a means of real-time asset valuation. With digitized assets, it would be straightforward to obtain immediate values for liquid assets. Illiquid assets could be valued different ways, applying generally accepted accounting principles to business inventory price data drawn from Amazon, enterprise eMarketplaces, and procurement networks. Smart contracts could consult websites such as LoopNet for commercial real estate prices. There would be many layers of private views for the different parties involved in reading and writing to the shared ledger. The point is that an organization's assets might start to have real-time valuations, which is a new kind of information affordance that could lead to reinventing business processes and have a significant benefit to firms, trading partners, and customers.

6.3 Payment Channels

A further consequence of digital assets is the possibility of new forms of financial interaction that make better use of business capital such as payment channels. Since assets are digitized, this means that they may be contractually obligated in ways that provide more assurance and trust to both owner and counterparty. Capital is tied up unproductively in the friction of conducting business, particularly international business. Ripple cites firms having $5 trillion in local cash balances in their countries of operation. $3.9 trillion of working capital is obligated in global supply chains [33]. There is an estimated $1.5 trillion global trade finance gap (i.e., trade finance transactions rejected by banks but needed for global distribution) [34]. Business requirements that have traditionally restricted the use of capital might be eased by Blockchain technology's ability to transfer payments immediately and instantiate ongoing credit relationships across borders. Further, Blockchains enable a lower cost of detailed control which allows new forms of remuneration structures such as payment channels. Digitized assets mean that it is easy to have "an account relationship" with business partners instantaneously because assets can be trustfully pledged on digital networks without having to know the other party (and verify them in an internal vendor/partner qualification process). Payment channels is the idea of contractually obligating an asset (a prepaid escrow of capital or another asset obligation), tracking consumption of a resource against the escrow, and then settling on a net basis in one transaction at the end of the period.

6.4 Business Networks and Shared Business Processes and Ledgers

Digitized assets and payment-channel type trading operations means that value chains may start to migrate toward single shared business processes for the conduct of operations. Shared business processes in business networks with privacy-specific views are the explicit objective of enterprise Blockchain implementations. For example, any party in a certain manufacturer's automotive supply chain might look up an item number in the Blockchain-based ID system using the same network-based process. In the implementation of Blockchain technologies, business practices might be redesigned and streamlined into shared processes in business networks. The implication of shared business processes is that there could also be shared ledgers which incorporate the economic side of business processes. In the farther future, instead of each firm maintaining its own books, it might have journal entry posting privileges for its activities, and a view of its overall activity in the shared ledger that keeps the books for all activity in the value chain. Firms might continue to run their own books until they fully validate and trust the industry shared ledger. The concept is "Ripple for ERP," a single shared set of business processes, accounting books, and legal processes in a value chain [6]. Blockchain implementations are underway in the financial sector for shared business processes and ledgers for securities settlement and clearing, syndicated loan placement, and interbank transfers. Blockchains are providing a means of improving the efficiency of existing operations, and also producing even more disruptive change that may force industries to reinvent themselves in new ways. One example of Blockchain technology enabling a new business paradigm is that of decentralized stock exchanges, as announced by tZERO [35].

In the farther future, integrated supply chain ledgers is the idea of having open permissioned links in corporate accounting software packages such that trade partners could post to each other's ledgers. Integrated ledgers might be hosted by Blockchain cloud services vendors or accounting firms. The implication of every business's books being kept on a distributed ledger is that they could be easily permissioned and pooled for single ledger views and net settling at the supply chain level. Net-cleared accounts might reduce supply chain debt and working capital commitments by 95%. Autoposted net-cleared transactions could automatically ripple across the entire supply chain.

6.5 Use Case: Global Supply Chain and Automotive Industry

The idea is that there would be a single shared set of processes in a business Blockchain, instantiated on a Blockchain network. All parties in a value chain would access the business Blockchain, having known and credentialed users. There would be private views and read–write access privileges assigned to certain parts of the Blockchain network. Let us consider an automobile manufacturing example. The manufacturer would create the new entries in the ledger. This would be Blockchain assets corresponding to physical world assets with a unique identification number (address in the Blockchain) in the Blockchain. Ownership would be assigned to the producer. There could be different steps in the manufacturing process as subassemblies and assemblies are made and shipper to the final assembly facility, all tracked in the business Blockchain. The finished product is shipped to distributors who also have private read and write access to certain operations in the share business Blockchain. The asset ownership is updated to reflect the distributor being the owner. The distributor sells the items, updating the ledger and transferring ownership and warranties to the customer. The customer owns and maintains the item, interacting with service and repair shops that log their activity to the same business Blockchain and have their own views. Customer applications might help the owners manage and predict maintenance, and have a lifecycle plan for their vehicle. When the customer sells the vehicle or it is ultimately recycled, these too are updates to the shared business ledger for ownership, disposition, taxes, insurance, and other consequences. The recycling plant repurposes the material and updates the ledger for their activity. Regulators and compliance bodies would have read–only access views to portions of the ledger, possibly for restricted time windows.

Business Blockchains such as the one just described include not only manufacturing, distribution, sales, and after–market processes but also ownership, financial, legal, and insurance processes that occur at every step and might be easily incorporated into the same Blockchain network as a platform for handling these diverse transactions between multiple parties. Now let us consider a real–life use case in an automotive Blockchain network of automotive industry recalls and counterfeit airbags. Airbag recalls are a costly liability for automanufacturers. In the first quarter of 2017, 40% of recalls by the National Highway Traffic Safety Administration in the United States were for airbags (9.6 million airbag recalls). Further, perhaps one third of airbags installed are counterfeit [36]. Repair shops buy "certified" manufacturer airbags from online suppliers, but do not have a way to validate their authenticity. The business case is that 30% of global airbags sold and installed are counterfeit. The solution is having a business Blockchain with a single shared process for airbag registration and lookup. Even before having full vehicle (VIN) lookup and registration on Blockchains, there could be component registration and lookup for items such as airbags.

Automotive applications are an important benchmark use case for Blockchain technologies in other ways too, particularly for the operation of commercial fleets, including in cold storage for food and vaccine delivery. It is estimated that there are 1.2 billion motor vehicles in operation worldwide in 2018 [37]. Of new cars sold, 35% in 2015 were Internet–connected and this is forecast to be 100% in 2020. Smart cities, connected cars, and autonomous driving initiatives are converging in secure vehicular gateway systems featuring Blockchain technology. The start–up StaTwig is such a cold storage Blockchain provider. Secure telematics operated by Blockchains deliver attestation, custody, insurance–tracking along with credential management, encryption, and digital signature technologies.

6.6 Use Case: Healthcare

In the future, there might be different kinds of Blockchains (ledgers) for different industries, for example, financial services Blockchains, supply chain Blockchains, and health industry Blockchains. Different industry Blockchains could record and track different kinds of processes, and exchange and provide access to different kinds of assets, including digital health assets. Blockchain health is the idea of using Blockchain technology for health-related applications. The key benefit behind Blockchain health is that the Blockchain provides a structure for storing health data on the Blockchain such that it can be analyzed but remains private, with an embedded economic layer to compensate data contribution and use. Personal health records could be stored in off-chain encrypted files stores, and called from pointers registered on Blockchains, essentially comprising a vast EMR system. Taking advantage of the pseudonymous (i.e., coded to a digital address, not a name) nature of Blockchain technology and its privacy (private key access only), personal health records could be encoded as digital assets and put on the Blockchain just as digital currency. Individuals could grant doctors, pharmacies, insurance companies, and other parties access to their health records as needed via their private key. In addition, services for putting EMRs onto the Blockchain could promote a universal format, helping to resolve the issue that even though most large health services providers have moved to an EMR system, they are widely divergent and not sharable or interoperable. The Blockchain could provide a universal exchangeable format and storage repository for EMRs at a population-wide scale.

One benefit of creating standardized EMR repositories is exactly that they are repositories: vast standardized databases of health information in a standardized format accessible to researchers. Thus far, nearly all health data stores have been in in accessible private silos—for example, data from one of the world's largest longitudinal health studies, the Framingham Heart Study. The Blockchain could provide a standardized secure mechanism for digitizing health data into health data commons, which could be made privately available to researchers. One example of this is DNA.bits, a start-up that encodes patient DNA records to the Blockchain and makes them available to researchers by private key.

6.7 Use Case: Digital Identity and Credentials Applications

Storing documents (home deeds, auto titles, wills) on Blockchain networks has been possible for years with services such as Storj.io and IPFS. This essentially a decentralized version of Dropbox or Google Drive. MIT has the open Digital Certificates Project program, issuing diplomas and academic credentials with a Blockchain ledger. A future application could be the norm to have a QR code at the top of a resume or CV which automates the academic credentials, professional certifications, work history, and criminal records checks that hiring companies perform in an expensive and manual way at present, confirming all of these data via Blockchain applications. Regarding identity credentials, the State of Illinois in the United States has been piloting a Blockchain-based birth registry system (with the Blockchain digital identity provider Evernym's underlying technology) starting in August 2017 [38].

6.8 Use Case: Higher Education and the Scientific Research Process

Far from being immune from the business process reinvention forced by Blockchains, universities are being required to reinvent themselves in the digital era too. At some point, national scientific funding agencies such as the NSF in the United States might launch a research Blockchain to manage grant proposal review and funding operations. The lengthy processes of preparing, submitting, and approving grants could be streamlined with digital identity confirmation and smart contract orchestration by Blockchains. In academia there is considerable interest in using Blockchain technologies in the context of higher education and research institutions. Blockchain technologies might be used to improve and reinvest how universities conduct their operations, and themselves be the object of study, both in scientific and humanities fields. For example, in 2017, NASA awarded a grant to the University of Akron for research into data analysis and to build a resilient networking system based in part on the Ethereum Blockchain for use in space exploration [39]. There are many research grants and foundation investors seeking academic study on the topic of Blockchain technologies and cryptocurrencies. The nature and concept of the university are being reinvented too, just as universities have founded Data Sciences Institutes in the past decade, and they are not also starting Blockchain Institutes.

6.9 Use Case: CryptoKitties and Primalbase

One of the biggest successes in Blockchain applications so far is Crypto-Kitties, a digital collectibles game in which users can buy, collect, "breed," and sell unique digital cats using Blockchain technology's ability to facilitate unique digital content. Cats breed by creating offspring with unique DNA preserved on Ethereum's Blockchain. The project is notable as a business success and also as a break-out new technology adoption case for attracting mainstream users to the platform. CryptoKitties launched in November 2017, and counts 250,000 CryptoKitties players, over 500,000

CryptoKitties sold for over $40 million [40]. The business model is that the project creator and operator, AxiomZen, takes a 3.75% transaction fee, and reaching more customers with a mobile roll-out in China.

One creative use of an asset token is the Primalbase Token (PBT) used in the Primalbase technology coworking office spaces in Amsterdam and Berlin. Purchasing one PBT allows the owner to control one seat in the coworking space. A token owner reserves a seat with an online interface that checks their token ownership and books their coworking seat. Four tokens correspond to an office at the coworking space. The PBT tokens are linked to Waves (the online playlist software), allowing the additional functionality to rent out your seat. The Waves functionality allows you to rent your playlist for a week or a month or other time frames, and with this linked to PBT, coworking seat owners can make their seat a liquid asset that they can monetize by renting it to other users, thus increasing the effective utilization of the resource. PBT is emblematic of the user-driven cryptoeconomy, where clever applications merge one or more current technologies.

7. CONCLUSION

This chapter discusses Blockchain distributed ledgers as one of the fastest-adopted technologies in human history, with 57% of large global entities considering or involved in their implementation. Blockchains are an important advance, but not an IT panacea. They are good in use cases involving secure real-time monetary asset and information transfer among multiple parties in a business network. There are many risks to consider with Blockchain technologies. Practitioners should always ask why using a Blockchain system would be better than using a centralized database for any particular business case. Over time, large entity IT infrastructure will likely include both centralized and decentralized elements in an overall system. Other risks are that Blockchain technologies are difficult to understand conceptually, technically, and operationally by consumers and business persons alike. Further, although the cryptosecurity has proven unbreachable thus far in the networks themselves, they are under constant attack at the endpoints and have been attacked here (the hacking scandals, exchange raids, wallet corruptions, and other problems are due to insecure and vulnerable system endpoints). These issues create barriers to the implementation and successful wide-spread use of Blockchain solutions.

Despite the inevitable risks, Blockchain solutions nevertheless present an exciting and inexorable step forward in instantiating the only sectors not yet updated for the Internet age, reinventing economics, finance, politics, and legal operations for the digital era. This chapter discusses Blockchain distributed ledgers in the context of public and private Blockchains, enterprise Blockchain deployments, and the role of Blockchains in next-generation artificial intelligence systems, notably deep learning Blockchains. Public Blockchains such as Bitcoin and Ethereum are *trustless* (human counterparties and intermediaries do not need to be trusted, just the software) and *permissionless* (open use), whereas private Blockchains are *trusted* and *permissioned*. Enterprise Blockchains are private (trusted, not trustless) immutable decentralized ledgers, with varying methods of reaching consensus (validating and recording transactions). Four enterprise systems are examined: R3's Corda, Ethereum Quorum, Hyperledger Fabric, and Ripple. New business analytics and data science methods are needed such as next-generation artificial intelligence solutions in the form of deep learning algorithms together with Blockchains. The hidden benefit of Blockchain for data analytics is its role in creating "clean data"; validated, trustable, interoperable, and standardized data. Business Blockchains may develop across industry supply chains with shared business logic and processes, and shared financial ledgers. Payment channels and smart contract asset pledging may allow net settlement across supply chains and reduce debt and working capital obligations. Specific use cases are considered in global automotive supply chains, healthcare, digital identity credentialing, higher education, digital collectibles (CryptoKitties), and asset tokens (Primalbase).

REFERENCES

[1] S. Nakamoto, Bitcoin: A Peer-to-Peer Electronic Cash System, 2008. https://bitcoin.org/bitcoin.pdf.
[2] R. Browne, Blockchain Technology Being Considered by More Than Half of Big Corporations, According to Study, CNBC, 2017.
[3] M. Swan, Blockchain: Blueprint for a New Economy, O'Reilly, Sebastopol, CA, 2015.
[4] M. Swan, P. de Filippi, Introduction, in: Toward a Philosophy of Blockchain, Metaphilosophy, vol. 48(5), Wiley & Sons, New York, 2017, pp. 603–619.
[5] M. Swan, Anticipating the economic benefits of Blockchain, Technol. Innov. Manag. Rev. 7 (10) (2017) 6–13.
[6] M. Swan, Blockchain economics: "Ripple for ERP" integrated supply chain ledgers to free $3.9 trillion in capital? Eur. Financ. Rev. (2018) 24–27. http://www.europeanfinancialreview.com/?p.21755.
[7] P. Tasca, Digital Currencies: Principles, Trends, Opportunities, and Risks, Bundesbank and ECUREX Technical Report, 2017.
[8] S. Roos, P. Moreno-Sanchez, A. Kate, I. Goldberg, Settling payments fast and private: efficient decentralized routing for path-based transactions, arXiv (2017). 1709.05748. Preprint appearing at NDSS 2018, 1–15. https://arxiv.org/pdf/1709.05748.pdf.
[9] Ethereum, Blockchain App Platform, 2018.

[10] A. Rosic, What is Ethereum? A Step-by-Step Beginners Guide, BlockGeeks, 2017.

[11] Eth, How Would I Explain Ethereum to a Non-technical Friend? StackExchange, 2018.

[12] M. Fang, Intro to Ethereum & Smart Contracts, Blockchain, Berkeley, 2018.

[13] J. Chow, Quirks With Smart Contract Programming, Ethereum Silicon Valley, 2017.

[14] V. Pradeep, Ethereum's Memory Hardness Explained, and the Road to Mining it With Custom Hardware, Vijay Pradeep blog, 2017. https://www.vijaypradeep.com/blog/2017-04-28-ethereums-memory-hardness-explained/.

[15] A. Hern, Bitcoin Mining Consumes More Electricity a Year Than Ireland, The Guardian, 2017.

[16] Accenture, Project Ubin Phase 2 Reimagining RTGS, 2017.

[17] Bank of Canada, Project Jasper White Paper, Payments Canada, Bank of Canada, and R3, 2017.

[18] Jpmorganchase, Quorum White Paper, JP Morgan Chase quorum-docs, 2016.

[19] W. Zhao, Banking Giants Send $30 Million in Securities Over DLT, Coindesk, 2018.

[20] N. Liao, On Settlement Finality and Distributed Ledger Technology, Notice and Comment, 2017. http://yalejreg.com/nc/on-settlement-finality-and-distributed-ledger-technology-by-nancy-liao/.

[21] S. Friedman, IRS Uses Tech to Track Bitcoin Transactions, GCN, 2017.

[22] P. Moreno-Sanchez, M.B. Zafar, A. Kate, Listening to whispers of ripple: linking wallets and deanonymizing transactions in the ripple network, Proc. Priv. Enhanc. Technol. 4 (2016) 436–453.

[23] A. Miller, M. Möser, K. Lee, A. Narayanan, An Empirical Analysis of Linkability in the Monero Blockchain, 2017. https://arxiv.org/abs/1704.04299.

[24] D. Chow, S. vs, Dynamic Credential Salting Explained, eForensics Magazine, 2017.

[25] D. Meyer, More Banks Are Trying Out Blockchains for Fund Transfers, Fortune, 2016.

[26] R. Zagone, Federal Reserve Task Force: Ripple Improves Speed and Transparency of Global Payments, Ripple, 2017.

[27] S. Marquer, Ripple's Blockchain Network Is Now More Than 100 Strong, Ripple, 2017.

[28] M. Dell Amico, Y. Roudier, in: A measurement of mixing time in social networks, STM 2009, 5th International Workshop on Security and Trust Management, September 24–25, 2009, Saint Malo, France (Saint Malo, FRANCE, 09 2009), 2009.

[29] OysterIMS, Dealing With Information Growth and Dark Data: Six Practical Steps, 2017.

[30] C. Crawford, An Introduction to Deep Learning, Algorithmia, 2016.

[31] N. De, Vermont City Pilots Land Registry Record With Blockchain Startup, Coindesk, 2018.

[32] J. Young, Sweden Officially Started Using Blockchain to Register Land and Properties, Coin Telegraph, 2017.

[33] PWC (PricewaterhouseCoopers), Bridging the Gap 2015 Annual Global Working Capital Survey, 2015.

[34] The Economist, Technology is Revolutionising Supply-Chain Finance, 2017.

[35] B. Dale, Investors Commit $100 Million to tZERO ICO, Coindesk, 2017.

[36] M. Ahlers, Feds Warn of Counterfeit Airbags Being Installed as Replacements, CNN, 2017.

[37] Statistica, Share of New Cars Sold That Are Connected to the Internet Worldwide From 2015 to 2025, Connected Cars: Statistics & Facts, 2018.

[38] M. del Castillo, Illinois Launches Blockchain Pilot to Digitize Birth Certificates, Coindesk, 2017.

[39] ETHnews, NASA Awards Grant for Ethereum Blockchain-Related Research, 2017.

[40] L. Barron, CryptoKitties Is Going Mobile. Can Ethereum Handle the Traffic? Fortune, 2018.

ABOUT THE AUTHOR

Melanie Swan is a Technology Theorist in the Philosophy Department at Purdue University. She is the founder of several startups including the Institute for Blockchain Studies, DIYgenomics, GroupPurchase, and the MS Futures Group. Ms. Swan's educational background includes an MBA in Finance from the Wharton School of the University of Pennsylvania, an MA in Philosophy from the New School for Social Research in New York, NY, an MA in Contemporary Continental Philosophy from the Centre for Research in Modern European Philosophy at Kingston University London and Université Paris 8, and a BA in French and Economics from Georgetown University. She is a faculty member at Singularity University and the University of the Commons, an Affiliate Scholar at the Institute for Ethics and Emerging Technologies, and an invited contributor to the Edge's Annual Essay Question.

CHAPTER SIX

The Use of Blockchains: Application-Driven Analysis of Applicability

Bruno Rodrigues, Thomas Bocek, Burkhard Stiller
Communication Systems Group (CSG), Department of Informatics (IfI), Universität Zürich (UZH), Zürich, Switzerland

Contents

Abstract

Bitcoin paved the way for a highly successful FinTech (Financial Technology) application on the basis of a newer technology termed blockchain. As this technology has evolved very recently beyond the financial market, gaining more public attention, other promising blockchain applications areas and use cases are emerging. However, it is still not a straightforward in which application areas and use cases blockchains do provide an advantage beyond distributed databases and decentralized systems. Through a decentralized and immutable data storage blockchains enable the enforcement and verification of exchange assets, while disrupting potentially existing business models by promoting the disintermediation of processes involving multiple stakeholders. While having no mediator or a third party controlling the operation, less operational costs and a higher business agility are targeted at. However, the adoption of blockchains requires not only overcoming its technical challenges, but it also implies to reach the ability to

Advances in Computers, Volume 111
ISSN 0065-2458
https://doi.org/10.1016/bs.adcom.2018.03.011

understand and handle the impact on existing business models, processes, and legal compliances. This chapter will overview the main blockchain application areas by providing examples and by discussing from the technical perspective on how and under which conditions the use of blockchains can be useful for a successful use case.

1. INTRODUCTION

While blockchains have started with FinTech applications [1], many other application areas, use cases, and specific blockchain types are emerging with the growth of public attention around the blockchain technology. The capacity to provide a trustworthy, decentralized, and publicly available data storage makes blockchains an interesting opportunity for organizations to increase business agility and reduce costs by removing intermediaries in distributed applications (i.e., by involving multiple and initially nontrusted stakeholders). It is precisely the powerful characteristic to disintermediate trust, which arouses interest in the use of blockchains in applications areas beyond FinTech, such as fraud detection, global access or ownership rights, contracting, and identity management [2,3]. It allows, for example, two parties to conduct an exchange upon an agreement without requiring the presence of a third-party acting as the (trusted) intermediary. Also, the ability to run program code managed and executed in a blockchain extends the possibilities for the development of new application areas on the blockchain [4].

Computer programs running within the blockchain are referred to as smart contracts (SCs), which define formal algorithms, following Truing-complete capabilities, comprising the agreements between stakeholders involved [5]. SCs are automatically enforced, verified, and replicated by nodes composing the blockchain to ensure its permanent storage and high obstacles to manipulate the contract's content. As SCs are executed on transaction data received by the blockchain, the contract is automatically enforced and records of this transaction are permanently added to the blockchain. When new blocks of records are appended to the chain, all other nodes are simultaneously updated, and there is no need for third-party intermediaries to verify or transfer ownership of these digital assets. Furthermore, the blockchain is agnostic concerning the type of data, which makes it applicable to a wide range of application areas where disintermediation of trust makes sense.

While short-term expectations of emerging technologies are usually overestimated, ignoring its long-term effects, the blockchain technology is foreseen as an evolutionary rather than revolutionary technology. The reason is that blockchains offer a radical change for a large variety of industries not only on the technical side but also disrupting existing business models, the operation of processes, and existing legal compliances. Shifting the *modus-operandi* of existing business models affecting its laws and regulations, regardless of its technology requirements, usually takes a longer time to be in place in a political sphere. Thus, it is important not only to combine the knowledge of a specific application area or use case with the blockchain knowledge but also to understand the long-term blockchain impact on existing business models, processes, and compliances specific to an application area to estimate the success of a case.

Foretelling the success of blockchains in a long term is not a straightforward task, in particular when considering that blockchains are still subject to active research and development by governments, industries, and academia. However, it is possible to identify key positive characteristics in its fundamental concepts that may influence a long-term adoption, such as disintermediation and automation that may lead to a reduction of costs. For the successful development of a blockchain-based application, one must also consider negative aspects that may be associated with business paradigm shifts and technical challenges that technology must overcome to establish itself in practice [6]. In this sense, based on blockchain fundamental concepts and characteristics, an analysis of existing application areas and use cases can facilitate the understanding of barriers of this technology's application and guide a successful long-term development and application of blockchain.

This chapter overviews the blockchain application areas providing examples of use cases and discussing how and under which conditions the use of blockchain can be useful for a successful use case. As usual with scientific work, references are utilized to refer to existing work, which, however, does not necessarily always be either complete or proven due to the young development stages of blockchains in general.

2. BLOCKCHAIN APPLICATION AREAS

In its purest form, a blockchain acts like a shared, replicated, append-only database where participants can depending on the type of blockchain, share write access, and participate in the validation process. While many blockchains implementations are typically designed to contain financial transactions (e.g., Bitcoin), it can also be designed to contain a general type of data (e.g., Ethereum). As a result, a complete overview of all application areas and use cases is impossible due to the variety of data that could potentially be stored, verified, and processed including the range of its application areas being as wide and as diverse as human activities themselves [7]. Thus, before investing in a blockchain start-up or using a blockchain product or service, it is extremely relevant to evaluate whether a blockchain application (i.e., use case) makes sense or not to avoid potential losses. Such discussion is conducted in Section 2.1 presenting arguments to guide the decision on whether to adopt blockchain or not and being fundamental to establish parameters to analyze blockchain application areas as well as their use cases.

2.1 Why Blockchain?

There is no general formula to determine whether a blockchain makes sense or not without a detailed analysis of the different blockchain properties and types, the requirements and goals of its potential participants, and the particular characteristics of the application itself. Each use case presents its complexity with different properties to be satisfied, and, similarly, each participant involved might have different requirements and goals. This decision-making problem has been discussed in Refs. [3,8] where the authors aim to determine whether a blockchain makes sense or not by defining a methodology or conditions that helps to answer why a blockchain is needed in a case-by-case analysis. For example, in Ref. [3] the authors present a flowchart to guide the decision-making process considering whether it is needed to store data for cases involving one or multiple entities that may not trust, or know each other. It is also considered whether a trusted third party exists and whether public verifiability is required among these entities to decide if a blockchain is necessary. The output of this decision process not only helps to decide if a blockchain is not needed at all but also helps to determine the type of blockchain to be used for a particular application in a positive case.

The authors in Ref. [3] state that a blockchain is not necessary if there is no need to store data or there is only a single entity involved in the use case. However, it is worth to note that even for a single entity that is large enough there might be a lack of trust between its departments or operations in different countries [8], and thus a necessity to validate its operations. Similarly, a blockchain is not recommended if it is possible to trust a third party or all involved parties are known and trusted to each other. For cases where a third party can be trusted, or there is no need at all for disintermediation, a blockchain-based solution might become a stepping-stone for data confidentiality and performance of transactions. Thus, when parties are known and trusted, a traditional database with shared access is likely to be the most suitable option to address these issues.

Conversely, for all cases that data needs to be stored and multiple entities that do not know to each other are involved, a permissionless (i.e., open and decentralized) blockchain is recommended when it is not possible to trust a central third party. In these cases, goods and services can be exchanged and verified publicly without a central entity managing the membership of these parties. Otherwise, a public-permissioned blockchain is necessary when public verifiability is required, but only some predefined parties can be trusted. For these cases, a central entity manages, the membership of participants, is responsible for granting or denying permissions to a set of participants to read or write in the blockchain. Lastly, a private blockchain is recommended for cases when data is stored by multiple known and trusted parties, and there is no need for public verifiability among them. For example, this type of blockchain could be deployed in a large company that seeks greater transparency and auditability in its internal processes, but do not want to expose these processes to general public.

Similarly, the author of Ref. [8] guides the decision process describing conditions, as in a checklist, that a particular use case or application should fulfill to justify the adoption of blockchains. Five (5) main conditions are described as essential, and three (3) conditions derived from these are described as recommendations. These conditions are [8]:

i. *Database*: need to store data.
ii. *Multiple writers*: blockchain is a shared database with multiple writers.
iii. *Absence of trust*: there needs to be a degree of mistrust between the multiple writers.
iv. *Disintermediation*: remove intermediaries by enabling databases with multiple nontrusting writers to be modified directly.
v. *Transaction interaction*: transactions created by different writers often depend on one other. For example, in transference of funds, allowing one to use the new funds.

Three (3) conditions derived from these are described as recommendations. These conditions are [8]:

i. *Set the rules*: the database should contain embedded rules restricting the transactions as the data recorded in the blockchain is not always reliable. Therefore, it is necessary to carefully design and control the order of data transfer to records in the blockchain.

ii. *Pick validators*: refers to the choice of trustable miners and the consensus protocol defined in the blockchain.

iii. *Back assets*: translation of nontangible assets to real-world assets.

As the authors in Ref. [3], the first two conditions described in Ref. [8] are straightforward concerning the need of a shared database, expressing, in the first place, the need to have a database (i) involving multiple writers (ii). Blockchains are, in principle, a shared structured repository of information and thus, there is a need to store data and have more than one party including information in this structure. However, there is the fact that most parties may not be known or trusted among themselves, which is described in the by their following condition. The capacity to provide trustworthy environment among nontrusted parties (iii) is one of the major blockchain characteristics, and cases involving one or multiple parties that requires a certain degree of trust, or do not trust each other at all, are essential to justify the adoption of the blockchain.

A blockchain application not only can provide trust over multiple nontrusted parties, but it also provides disintermediation (iv) enabling data to be directly shared across boundaries of trust in case there is no need for or absence of trust in the central third party. For cases which a central authority is not required to validate transactions between one or more parties, transactions can be independently verified and processed by every node with the blockchain acting as a consensus mechanism to ensure these nodes are synchronized. However, removing an intermediary might present drawbacks regarding performance and data confidentiality, and the adoption decision is a critical factor to the success of a use case. Some organizations have earned their trust over the years, and for use cases where confidentiality or performances are key factors, a trusted third party is naturally the best choice.

The next key condition described in Ref. [8] is the transaction interaction (v), which is similar to the public verifiability described in Ref. [3]. The author state that the true value of blockchain rises when there are interactions between transactions created by multiple writers and these transactions depend on each other. For example, in a supply chain process involving multiple stakeholders, a final party can only deliver goods or services when all transactions in the supply chain were submitted and validated, or a simple transference of funds between two parties, allowing one of these parties to use the new funds. This means that transactions submitted by nontrusted parties can be validated without exposing themselves to risk, allowing the delivery vs payment settlement to be safely performed over the blockchain without a trusted intermediary. While transaction interaction is similar to public verifiability, the author in Ref. [8] describes this condition as fundamental toward the blockchain adoption in a use case and authors of Ref. [3] state this requirement as a conditional to decide between public, in case verifiability is required, and private-permissioned blockchains otherwise.

The last conditions are derived from the previous ones and seen as recommendations rather than mandatory for the blockchain adoption. Thus, according to the authors in Ref. [8] when a use case fulfills the previous conditions is necessary to set the rules, select the validators, and back the assets in the blockchain. Set the rules is a consequence of the previous five conditions that are related to embedding rules restricting the transactions in the blockchain. These rules are fundamentally the logic behind the use case being typically described using SCs in a blockchain application. Next, is necessary to select the validators, which refers to the choice of miners and the consensus protocol defined in the blockchain to avoid conflict between transactions and thus multiple chains, transaction censorship, and biased conflict resolution. The choice of the consensus protocol is closely related to the needs of a particular use case and the involved parties, whether it requires transparency and disintermediation, or privacy and performance, among other requirements. Lastly, back the assets (viii) is related to the translation of a nontangible asset to a real world, physical asset, e.g., what is the ballast of some virtual tokens defined in a blockchain converted into some assets in the real world. This concern stems from the nature of the blockchain environment typically composed by multiple nontrusted parties that may attempt to create digital assets to obtain physical assets maliciously.

As mentioned earlier, there is no general formula to determine whether a blockchain solution makes sense or not for a particular use case. However, based on Refs. [3,8] it is possible to draw essential characteristics that a use case should fulfill to decide whether to adopt blockchain and make efficient use of blockchains. Furthermore, application requirements such as transparency, confidentiality, performance, fault tolerance, and disintermediation should be seriously accounted onto such decision. Thus, main requirements to adopt a blockchain, regardless of defining its characteristics, are summarized herein as follows:

- *Database*: first and foremost, to adopt a blockchain solution should exist the need to store data in a structured fashion.
- *Multiple writers*: may or not involve multiple parties, but should include multiple writers that may not know or trust each other.
- *Disintermediation*: required when a trusted third party is not available, or it is not needed nor desired for a particular use case.

- *Sequential transactions*: condition for uses cases where the order of transactions impacts further transactions, i.e., data is sequentially concatenated. The transaction model can be either UTXO (unspent transaction output) or SC based. Whereas the former is used in Bitcoin [1] and suited for the transfer and tracking of assets, the latter is used in Ethereum [4], and arbitrary logic can be defined, executed, and verified.

It is mandatory that these conditions should be fulfilled (i.e., a use case should need these characteristics) to answer if a blockchain solution is ideal or not for a particular use case. If one or more conditions are not satisfied, then another type of data storage structure should be considered. However, this is just an indicator that blockchain can be a good solution to the use case, case all conditions mentioned earlier are satisfied, it is also necessary to discuss the use case requirements and the trade-offs between them. These requirements are summarized as follows:

- *Performance*: blockchains are naturally slower than any centralized database. Besides of the regular operations performed in a traditional database, a blockchain should perform extra procedures to ensure that transactions are correctly recognized, propagated, and verified. Also, traditional databases have been deployed, tested, and optimized through decades of development.
- *Distributed or centralized control*: it is closely related to the disintermediation blockchain characteristic. The question to be answered is whether the use case requires to be centrally managed or not. The core value of blockchain stems from its power of disintermediation, ensuring that nontrustable parties be able to interact in a trustworthy environment.
- *Privacy and confidentiality*: users and data in a blockchain are visible to all nodes (with reading permission) participant in the blockchain. For use cases where a user and data transparency are desired, blockchain is the choice to verify and process every transaction transparently independently. However, if privacy and confidentiality are required, and there is a trusted third party, a blockchain poses no advantage over a centralized database. However, it is still possible to hide confidential information on a blockchain, but hiding critical information requires lots of cryptographic efforts for the nodes in the network.
- *Robustness or reliability*: it is related to the dependability of the database. Blockchains are typically fault tolerant due to its built-in redundancy, which means that transactions are replicated within the blockchain network and permanently recorded. Some trusted third parties can also provide fault tolerance offering replication techniques, but blockchains can provide high redundancy by default.

Not only these use case requirements should be evaluated solely by its own to decide which type of database is necessary, but also the trade-offs between them. For example, as mentioned in Ref. [8], if disintermediation and robustness are more important than performance and confidentiality, then a blockchain-based solution seems to be the natural choice for the use case. Alternatively, if performance and confidentiality are required then other types of data storage such as a centralized database or a distributed master–slave replication structure might be more adequate.

Although confidentiality is typically the most relevant criteria and despite some attempts to provide privacy in blockchains (e.g., zero-knowledge proofs, side channels, and encryption), blockchains are not meant to store sensitive information, and thus off-chain centralized solutions can be combined with the blockchain network to fill this gap. Furthermore, any combination of these requirements is valid and should be carefully considered. For instance, taking requirements which are not always on the same side of the road in a single storage solution, such as performance and robustness or confidentiality and distributed control, it may not be straightforward to determine which whether a blockchain is necessary. Thus, in these cases is not only essential to observe the use case requirements to decide whether to use blockchain but also to check whether the use case complies with the blockchain characteristics.

Also, answering these trade-offs is helpful to guide the decision on which type of blockchain to deploy, in addition to the steps discussed in Refs. [3,8]. A permissionless blockchain is a fully decentralized approach removing any intermediary certifying transactions between parties, and it is a determinant factor is when the participants are not known or trusted to each other. Possible downsides for use cases implemented in a permissionless blockchain are performance and confidentiality once all data in the chain is open and publicly verifiable, and it requires a consensus mechanism to maintain the chain order. When participants are known but not trusted, write permissions may be distributed according to trust criteria in a public-permissioned blockchain (i.e., consortium) and data within the chain could be verified by all members. The difference between a public and a private-permissioned blockchain is that read and write permissions are managed centralized, avoiding public readability and increasing the blockchain confidentiality. However, it is commonly argued [9] if a private-permissioned blockchain is merely a traditional centralized system with a degree of cryptographic auditability attached with no reasons to belief that cryptographic authentication based on hash-chained blocks containing Merkle tree roots is an optimal format of audibility. Also, it does not prevent single (large) entities managing read and write permissions to perform fraudulent actions on the data by freely changing rules.

In summary, it is not only difficult to determine whether a blockchain infrastructure is necessary or not but also and in case it is needed, which type of blockchain to be deployed. A decision should be based on the features offered by each type of blockchain (and conceptual differences among traditional databases schemes and blockchain), and application requirements and its trade-offs.

2.2 Application Areas and Use Cases

Blockchain moves the control of daily interactions from centralized entities, redistributing it among users in a decentralized and transparent way. The ability to remove intermediaries creating a secure bridge between two parties is a factor that modifies how some use cases are currently made. However, the ability to operate over any data makes that extends its application areas making it as diverse as the human activities themselves—as long as some conditions were observed to ensure its necessity and efficiency. Still, one of the most useful ways to understand and determine whether blockchains are been able to improve the efficiency of a use case is observing its successful application areas and use cases. Therefore, this section presents selected blockchain application areas and use cases, analyzing the reasons for and against the use of blockchain in these areas.

2.2.1 Supply Chain

Goods all over the world are produced and distributed across a vast network of producers, retailers, distributors, transporters, and vendors in a complex arrangement of processes. The availability of standardization and certification bodies has done a remarkable work improving the producers quality and consumers awareness, but actual processes remain costly and unreliable, especially in regions with high levels of corruption, as reported in Ref. [10]. Blockchains counter these shortcomings providing:

- *Transparency*: every asset inserted in the blockchain can be tracked and traced by all participants of the supply chain.
- *Trust*: since the whole system is running transparently and every transaction is signed and traced back to its owner, it is easier to check malicious behaviors. Also, there is no need to trust a third party which could manipulate the data.

- *Interoperability*: single, transparent, and trusted source of data that is shared between different systems in the supply chain. Therefore, instead of many systems individually managing databases, the visibility of data is limited.
- *Reduced costs*: costs associated with all kinds of manual paperwork, legal documentation processing, duplicate entries, and mismatches can be reduced with digital processes.

While costs can be reduced by removing intermediaries and costs associated with legal documentation processing, risks can be eliminated by the inherent data transparency and the use of cryptographic primitives to ensure ownership of each step registered in the supply chain ecosystem. For example, when something goes wrong within a vehicle or aircraft that are systems composed of various subsystems, it is important to know the origin of each subsystem to collect details like manufacturer, production date, how it was transported, and other relevant pieces of information. Thus, it is essential not only that information is available to all participants, but also that no central authority owns the data of multiple steps in the supply chain, so as its origin can be tracked to a single entity.

The author of Ref. [11] states that the scale and complexity of the involved systems often lead to high transactional costs, mismatches, and errors in manual paperwork. Thus, digitalizing information and making it available to all members of the supply chain can optimize the time-to-market of products and increasing end-to-end transparency. Therefore, with a blockchain-based solution reducing the amount manual paperwork, which is more prone to error than digital documents, everyone has access to the same, up-to-date information. Furthermore, blockchain can be used not only for registering, certifying, and tracking goods but also to trigger automatic payment systems and other procedures through SCs. However, as detailed in Ref. [12], as the entire process typically involves many entities with different organizational processes, well-designed interfaces and data-models should be defined to enable this transfer of information to occur smoothly.

Although reduced time to market is a common objective, performance requirements are not as relevant as other features like transparency, trust. It is not important whether a supplier is fast if the goods are not registered or delivered correctly. Thus, a reliable and transparent supplier is more important to a wholesaler than a fast one. Also, while disintermediation is naturally obtained by deploying a blockchain solution, transparency is achieved by defining a public or private-permissioned blockchain, depending on the use case. For example, reading permissions of internal steps of a supply chain may not be exposed to general public (e.g., industrial confidentiality), but only to its internal members to increase the overall process efficiency. Similarly, writing permissions shall be given only to those who take part in the system.

Notable use cases stem from the logistics sector, responsible for the transport of goods both overseas and land. For example, IBM and Maersk [13] announced in 2017 the creation of a network of shippers, freight forwarders, ocean carriers, ports, and customs authorities. The goal is to reduce the costs associated with paper-based trade documentation processing and administration, which are estimated to be up to one-fifth the actual physical transportation costs. In 2014, Maersk found that just a simple shipment of refrigerated goods from East Africa to Europe can go through nearly 30 people and organizations, including several interactions and communications, as reported in Ref. [13]. The solution, based on the IBM Hyperledger blockchain, is intended to enable real-time exchange of events (e.g., when a good is shipped or received) and documents (e.g., trade reports) reducing errors and delays in the process. For authorities, it will also provide traceability as goods advance and enters in countries for border inspection as the solution can give real-time visibility, significantly improving the information available for risk analysis and targeting, eventually leading to increased safety and security as well as greater efficiency.

Other relevant use cases appear in the food supply chain, which had seen a big change over the years with the fast increase of population in urban areas, with more than half of the world's population living in these areas [14]. As a result of the need to provide more food in a decreasing time, more cases of fraud appear through the entire food chain, being not enough to avoid fraud even if each country has an organization body defining minimal sanitary conditions for production, transport, storage, and sales of food. These cases directly affect not only the health of consumers, but it also creates economical and trusts losses throughout the entire supply chain. Among the main requirements to avoid frauds in the food sector are [14]: (i) traceability of food producers and transporters; (ii) transparency to track and trace the parties involved in the supply of fragmented foods (e.g., hamburger); and (iii) trust among entities. Also, even if a blockchain-based solution does not solve all fraudulent actions in the sector, it does comply with these main requirements.

An example comes from Switzerland focused on the agricultural sector, whereas the Swiss Federal Constitution provides incentives in the form of subsidies to encourage farmers whose productions comply with environmental and animal-friendly regulations. Based on this, the University of Zurich developed a blockchain-based system to demonstrate how such system can increase transparency and automate interactions in the agricultural sector [15]. Once a farmer creates a new request for *Direktzahlungen* (subsidy paid by the government), the farm has to be inspected and periodically checked. Then, a SC is deployed to the blockchain containing all relevant information such as the type of animal and their living conditions to calculate the subsidy amount. These conditions are verified by government agents during inspection visits on the farms. Then, the payout is accomplished through a transfer of ERC20[a] tokens to the farmers. However, the downside of the system requires an exchange platform which guarantees that farmers can exchange their tokens into real currency.

Another use case appears in the pharmaceutical supply chain with the transport of medicinal products for human use. The new European Union regulation for distribution of medicinal products, GDP 2013/C 343/01 [16], states that it is an obligation to report any deviation on the transport (such as temperature and humidity) as well as to monitor parcels at all times, to the distributor and the reception of these products. In this regard, a modum.io AG developed a blockchain-based solution combined with IoTs (Internet of Things) sensors to monitor all necessary data during transport [2]. From a business perspective, blockchain allows reducing the number of intermediaries and SCs can ensure compliance with GDP regulations verifying data collected by IoT sensors during transport.

From a practical point of view, valuable lessons were presented in Ref. [2], which are summarized as:

- *Connectivity*: most warehouses have poor or no Internet connectivity which has an impact on the design of sensors (to store data offline) and to create SCs to register a new shipment.
- *Simplicity*: a key in all aspects of the system design because, in most cases, the solution will replace manual paper-based documentation, and workers do not have the technical knowledge to operate devices and digital processes. The use of blockchain has to be as transparent as possible to a system user.

In a nutshell, blockchain has the potential to be game changing the in the supply chain sector for several reasons that ranges from the digitalization and automation of process to the increase of trust in its members.

---

[a] ERC20 standard: https://theethereum.wiki/w/index.php/ERC20_Token_Standard.

However, even if data transparency in every step of the supply chain is a desired feature, this information should not be publicly accessible due to some industrial reasons. Thus, reading and writing permissions should be carefully managed in a permissioned type of blockchain. Also, as discussed in Ref. [2], many places and workers are still not ready for digitalization of processes, with many warehouses with poor or without connectivity and workers without knowledge to operate these processes. Lastly, it is important to consider that humans are the source of information and, as an immutable database, erroneous entries in the chain will be a source of inconvenience for the other participants.

2.2.2 Health Care Industry

Not only the supply chain is a difficult task for the health care industry, but it is an extremely complex system on its own. In 2015, the Office of the National Coordinator for Health Information Technology (ONC) published the Shared Nationwide Interoperability Roadmap [17], defining a path to improve interoperability for the next 10 years. Besides of recognizing that many advances were made in precision medicine, telehealth (the use of telecommunications and virtual technology to deliver health care outside of traditional facilities), and adoption of electronic health records and big data analytics, there is still progress to be made. The ONC established a roadmap to improve interoperability between different sectors and entities of the industry, which includes standards for identity, authorization and access data formats, data transmission, longitudinal health information, and provider workflows.

The industry has multiple interconnected entities which not always act in predictable ways to deliver health care services to patients. Actions and duties of these entities are regulated by governments and standardization bodies, such as the ONC, to ensure that effective patient care becomes the focus of health care delivery. However, due to the complexity of the sector, some key challenges are found [17,18]: (i) number of fragmented data maintained individually by these entities; (ii) patient data security; (iii) system interoperability; (iv) cost and time effectiveness. Blockchain can help to overcome most, if not all these challenges, depending on the use case-specific requirements. Also, existing health care systems do not need to be completely rebuilt to use a blockchain solution, being necessary only the way how these systems interact. The main benefits of blockchain in the sector are:

- *Transparency*: in general, the way how data is shared across the health care network is extremely important and using blockchain, it can eliminate duplicate entries, errors, and fragmented data that can happen in an individually managed database.

- *Trust*: this single view on shared medical records and increase trust and interoperability among these systems. However, it should be noted that patient records cannot be fully exchanged or exposed in the blockchain and, therefore, read and write permissions need to be carefully managed.

SCs could create an efficient method for accessing and analyzing patient records that can be permissioned to selected entities. In the end, by reducing inconsistency, transaction costs, eliminating third-party applications, and providing a single view of patient data selectively shared across the chain, can increase the cost-effectiveness in the health care industry. Several initiatives in and between the private and public sectors have been undertaken to develop blockchain-based solutions for the health care industry. One use case is MedRec [19], a project between MIT Media Lab and Beth Israel Deaconess Medical Center. The use of blockchain in this scenario is intended to give patients the ability to decide who can access their health care data and use these permissions to create an automated approach to data sharing for clinical and research use. The solution itself does not store patients' health records, it stores the record's signature (i.e., a hash of the data) on a blockchain, and this signature may assure that data's unaltered copy is obtained.

While the permissions, data storage location, and audit logs are maintained in the blockchain, all health care information remains in health records systems and requires additional software components to enable interoperability [18]. The project has been tested as a concept prototype with medication data, and the developers are looking to enhance the project's scope by adding more data types, data contributors, and users. As shown by this proof of concept, biomedical and outcomes research may significantly benefit from the application of blockchain to provide fast, secure access to research data.

A second use case was developed by Guardtime [20], a data security company that uses blockchain technology to promote public and private data security, which in partnership with the government of Estonia, created a blockchain-based framework to validate patient health records. All citizens have issued a smart card, which links their health records with their blockchain-based identity. Therefore, all updates in these health records are seen as a regular transaction with an assigned hash and registered in the blockchain framework. This approach ensures that data within the health records contains an immutable audit trail and that records cannot be maliciously modified. For example, in an appointment scheduling, it is assigned a time stamp and cryptographically signed in a block and these records could be verifiable using the blockchain guaranteeing system, process, and operational integrity.

The health care industry is a complex sector with several interconnected entities maintaining similar records of patient data in individually managed databases. Besides of the lack of interoperability, this creates a high-level fragmentation and mismatches between of patient data. Blockchain can significantly simplify interoperability among these entities as long as patients provide permission of its data to be shared, as shown in MedRec [19]. However, efforts from standardization bodies and perhaps, government's incentives, are also required to create a common data model and APIs so as the different systems from these entities can interact over the blockchain (a concern also outlined by ONC in its roadmap for interoperability [17]). In this regard, the government of Estonia gave the first step toward the blockchain adoption [20], creating a nationwide system integrating data from Estonia's different health care providers to create a common record that every patient can access online.

It is important to note, however, that interoperability cannot be solely achieved only through a blockchain solution where all entities share data. Even if transparency is a desired feature to avoid fragmented data, there are different companies competing in similar areas (e.g., multiple health insurance companies) which may not wish to store information in a blockchain where competitors may read. Therefore, confidentiality is one of the main requirements that have to be addressed not only to secure private patient data but also to protect information that could provide a competitive advantage. An alternative in this sense is that companies could share a hash of data in the blockchain, and this hash could be verified only by an authorized entity (e.g., government or patient), but not to other competitors. Ultimately, the future will tell if all or only parts of that blockchain technically available characteristics and advantages can be practically exploited in the health care industry.

2.2.3 Internet of Things

IoT is an interconnected network of devices ranging from small sensors and actuators to vehicles and other large appliances, stationary or portable, that can automatically transfer data without requiring human intervention [21]. In this regard, these devices can be deployed as an enabler in a broad range of blockchain-based use cases, such as smart vehicles, tracking temperature during transport of medicinal [2] or food shipments [15], or sense patient data [19]. As detailed in Ref. [22], both manufacturers and customers can benefit from this combination. Manufacturers can reduce their maintenance costs by, for example, deploying a public-permissioned blockchain network of its devices to identify and maintain devices. Thus, whenever a new system update is released, identified devices can retrieve, from a predefined SC address, a hash-based link and download the update.

The SC can use, for instance, a distributed file system, such as IPFS[b] to distribute these binaries over the network and, instead of using a central system to perform updates on several devices, these binaries can be spread over the network automating and optimizing system's update tasks. From a customer perspective, it is also important to have devices registered in the blockchain. It means that messages signed by registered devices can have its authenticity confirmed, and these devices can interact through SC functions which then model the agreement between the parties. Therefore, not only a trustworthy device-to-device model can operate transparently but also in an autonomous way without a central authority.

The blockchain also provides a financial layer that can be used to create a marketplace of services between devices, as outlined in Ref. [22]. For example, a smart fridge could sense its interior and automatically reorder new items whenever required. Also, in the case above where devices can serve a copy of binaries for system updates, storage space, and API calls can be monetized to sustain infrastructure costs (e.g., storage, bandwidth). Therefore, every device or group of devices behind a sink can use cryptocurrencies accounts to expose its resources to other devices or users in an "as-a-service" basis and get a financial reward for it. All these interactions can happen automatically, without human intervention.

As IoT devices can be used to sense, store, and process data in many different environments, combining blockchain and IoT creates a vast possibilities of use cases. The combination IoT and blockchain can provide the following benefits:

- *Automation*: blockchain and SCs help to optimize and automate business processes through IoT.
- *Marketplace*: blockchain facilitates the sharing of services and resources leading to the creation of a marketplace of services between devices.
- *Trust and transparency*: greater levels of trust and transparency are provided by blockchains. While devices can automatically sense and process data for a particular use case, blockchains and SCs can provide a transparent and verifiable communication.

---

[b] IPFS: https://ipfs.io.

However, as there are different types of blockchains and devices, it is necessary to observe which combination of blockchain and devices is better suited to a use case. For cases where a public blockchain (and proof-of-work based) is required, the following challenges need to be addressed [21]:

- *Resources restriction*: mining is expensive, and the majority of devices is resource constrained.
- *Performance*: mining blocks consume time, while for most cases, low latency is desired.
- *Scalability*: blockchain scales poorly as the network size increases.
- *Security*: blockchain can create significant overhead traffic, which may be undesired for most constrained devices. Additionally, reaching confidentiality is extremely difficult since the content of every transaction is exposed to every blockchain node and methods to that allows confidential transactions (e.g., zero-knowledge proof, homomorphic encryption) are resource intensive and thus, not applicable to resource-constrained devices.

In private networks where participants are known but not trusted, a permissioned blockchain can be used with other protocols to reach a consensus which requires fewer CPU computations. This would reduce or eliminate the need for powerful devices, but privacy (i.e., identifying a device) and confidentiality (i.e., identifying the data) would still not be straightforward to be reached. However, it is worth to note that despite these considerations, the combination of blockchain technology and IoT devices can be very powerful to automate time-consuming tasks in a trustworthy and verifiable manner, and potentially creating a new ecosystem where devices can seamlessly communicate with each other and offer services.

Ultimately, blockchain and IoT are part of a revolution that modifies how platforms are currently defined and used, passing from technical infrastructure to an ecosystem enabler, and creating the foundation for new business models. Therefore, entities within this ecosystem should not only be aware of the technical challenges of this combination but also to redefine their business strategies to build cost-efficient solutions that take into account autonomy and privacy. On a business level, it is needed to redesign processes and agreements around traditional definitions of data and to define legal responsibilities and actions.

2.2.4 Identity Management

Digital identity is a growing issue on the Internet not only due to an increasing number of users and devices but also because of the user complexity to create, use, store, and verify these identities in many services fragmented across traditional centralized databases. This results in insecure systems where people often use the same combination of username and password for many services [23]. Also, the security of these identities is just as strong as the security of the database where they are stored, which makes large identity providers like Google and Facebook a valuable target for hackers to attempt to crack. Also, information taken from this large pool of identities can be a valuable source of information for market-oriented publicity.

Blockchain-based identity systems reduce this risk by decentralizing the ownership of data and bringing the responsibility of managing identities back to the user. Thus, in addition to the password security being the sole responsibility of the user, it becomes less economically valuable for hackers to crack many individual identities one by one. Also, the use of SCs may include sophisticated logic that helps with key verification, revocation, and recovery, lessening the key management burden for the end user. In summary, three main benefits could be observed in a blockchain-based identity service [23,24]:

- *Data ownership*: being the digital identity a reflection of a real physical person, the attributes of identity, or group of identities becomes valuable information for market-oriented publicity. In a blockchain-based identity, the information is decentralized, and users are the owners of their data being able to monetize according to its will.
- *Reduced risk of data breach*: well-established public–private key pair schemes (e.g., Elliptic Curve Digital Signature Algorithm—ECDSA) are commonly used to represent identities. Also, it becomes difficult for hackers to steal a large number of identities on a peer-to-peer basis.
- *Immutability and transparency*: blockchain algorithms and SCs are open and transparent, being independent of a single entity to manage and update. Moreover, it enables quick auditing of identities through the immutable chain.

However, besides these positive characteristics, it is necessary to evaluate the requirements of each use case since the identity service can, in most cases, provide access to a critical service to the user (e.g., online banking). Also, as pointed by the author in Ref. [25], some aspects are not favorable of the blockchain adoption:

- *Immutability myth*: while most authors state that blockchain is immutable, others say that data immutability is not a guaranteed condition. The immutability discussion refers to the 51% attack, where an actor, which controls 51% or more of the mining power of the network, would have enough economic power to reach more mining capacity than the remainder of the network and to subvert the network. Therefore, immutability is based on the assumption that is economically difficult to undermine the network, but for some entities (e.g., governments), this cost is still not high enough.

- *Lack of trust*: related with the lack of confidence (which, in its turn, is associated with immutability myth) and knowledge of individuals and companies in new technology and the absence of a legal framework to control liability are aspects that affect the trust in blockchain-based identities.
- *Synthetic identity and identity verification*: while synthetic identity is related to creating real and unique identities, the verification involves checking whether the user claiming an identity is a rightful owner.

Despite the author in Ref. [25] describes synthetic identity and identity verification as blockchain-related issues, these issues are not only related to blockchains but to all digital identity services. Similarly, the author describes a demographic issue, which is related to the problem of issuing strong digital credentials at scale for different groups of people. However, these aspects are related to the design decisions and how an application layer, that uses the blockchain as a storage back end, is implemented and not the blockchain itself. In this regard, many companies are investing in blockchain-based solutions to authenticate their users. Blockstack [26], for example, provides identities, naming, storage, and authentication services for decentralized apps using the Bitcoin blockchain as a back end to secure identity data [1]. The solution is built following three main principles: (i) decentralized naming and discovery service where users register and discover resources in a human-readable way; (ii) decentralized storage where users place their identity without revealing to the third party; and (iii) performance comparable to a traditional centralized service.

Blockstack contains a network of nodes with the global registry of identities, public keys, and other metadata, and operations are represented in the Bitcoin blockchain to provide a name registration service binding public keys to their names. These names are resolved into a new DNS (domain name system) built on top of the Bitcoin layer and every time a Blockstack node joins the Bitcoin network, it can synchronize its registry via the Bitcoin blockchain. For data storage, a DHT (distributed hash table) is used to store key-value pairs, where the integrity of the values can be verified in the Bitcoin blockchain. The solution not only simplifies the maintenance of the keys using a DNS service but also decentralizes the identities using the Bitcoin blockchain as back end.

Another blockchain-based digital identity is VALID [24], which based on a new European privacy regulation called the General Data Protection Regulation (GDPR), aims to connect people as physical identity owners with data consumers, such as product and service providers, brands, researchers, and advertisers [24]. The goal of VALID is to let the users be responsible for their data by granting permissions or monetizing their data through a marketplace. Thus, the company offers a platform consisting of a mobile wallet and a web application for users to manage their data and permissions, and a blockchain-based marketplace for consumers to buy and access consumer's data. Then, agreements between consumers and users are defined via SCs in the marketplace back end.

The company provides an example of how to take advantage of a new regulatory incentive to combine with new technology. Blockchains are used as a platform to record the trade of user data, where the user may sell the use of its data based on SCs, and to store a user data. Thus, a public-permissioned blockchain is combined with a permissionless blockchain, to tracking user data modifications of the former and recording these changes on the latter. Therefore, the permissionless blockchain acts as a notary for data modifications by checking whether and when they occurred. Ultimately, when users are the sole owners of their data on purchases, online browsing history, or mobile data, they can choose whether or not to monetize their data.

There is no doubt that Blockchain has a fundamental impact on existing digital identity methods, bringing back to the user the ownership and of digital identities data by centralized entities. However, some aspects need to be considered before adopting a blockchain-based identity service. The use of new technology based on cryptographic primitives may be less unintuitive for most people, impacting the adoption of blockchain-based services. Also, the performance is a restricting factor and, as in any fully distributed network, it is hard to tell which nodes contain what information, so the search has to go through the whole network to find specific data. Also, the more nodes are added to the network; the more difficult the information lookup becomes because it is unclear which nodes contain what information.

2.2.5 Government Services

Governments strive to be as open, transparent, and collaborative as possible, creating projects and actions that aim to promote transparency, fight against corruption, and increase social participation through the development of new technologies. A significant advantage for blockchains in public services stems from the disintermediation, which uses the decentralization of information as a safety measure to increase transparency and avoid frauds. Thus, information about finances, people, real estate, certificates, and other documents can be not only virtually unchanged once stored on the blockchain but also transparent and accessible to everyone. Also, blockchains can be used as a governmental platform to increase interoperability between the different governmental bodies, for example, reducing the risk of fraud or irregularities in the payment of social benefits.

The most prominent example of an open and transparent digitalized government stems from Estonia, which is a pioneer country in the adoption of new technologies by both the market and the government. The "e-Estonia" movement reflects a trend of modernization and adoption of technology by the Estonian government since the early 1990s, and in the last years, Estonia has offered a wide range of blockchain-based services, being the most famous the e-Residency [27]. Created in 2014, the system allows the creation of a unique digital identity that gives them access to various Estonian public services, including the e-Health service [20]. For example, Estonian citizens can log into their e-Health service using their e-Residency identity, which is verified in the blockchain, and then check their medical records and which medical professionals have done the same and when.

However, it is important to understand how these public services are connected with the blockchain to obtain its benefits fully. In Ref. [28], the author describes how the Estonian services are implemented and connected to the blockchain. The blockchain uses the Keyless Signature Infrastructure (KSI) from Guardtime [20], which generates and maintains the blockchain containing the distributed ledger. The KSI blockchain has been integrated into key government registries, such as the e-Residency, and other services, such as business registry, property registry, and official announcements, being are used for both internal and external processes to provide a signature service [28]. Also, it ensures the integrity of data enabling the efficient detection of both intentional and unintentional modifications of data because records and associated metadata are chained to previous records. Therefore, when the customer transmits the asset's hash (e.g., the hash of medical document), a token is received in return, which proves the existence of the document.

The system interacting with the blockchain is named X-Road. It acts as an interoperability platform connecting different governmental institutions to multiple databases. Thus, the role of the KSI blockchain in the Estonian government is to ensure the integrity of data recorded in the X-Road by providing a chronological proof-of-existence mechanism, which ensures the detection of changes in these registries [28]. Also, it enables an audit log (or transaction log) containing entries such as a time stamp, user login information, and accessed and modified resources are written to the KSI blockchain together with the hashes of the database records. As only hashes are stored in the KSI blockchain, there is a preservation of citizen's privacy, and it enables the transparent auditability and makes records impossible to delete without detection.

Another promising use case for blockchains in government services is electronic voting. In many countries, votes are manually cast, stored, counted, and verified by a central authority. This process is prone to error due to the manual verification of votes, and there are no reliable ways of auditing a completed election because it is done via a central authority. A blockchain-based voting would empower voters to do these tasks themselves, by allowing them to hold a copy of the voting record and recount votes if required [29]. However, there is the challenge to solve regarding the secrecy of the vote, which is related to the coercion of voters [3]. Thus, it is necessary to inhibit voter coercion procedures by third parties and the marketing of votes. In this case, pseudoanonymity is often considered a crucial requirement in the voting process, meaning that it should not be possible to track and prove the relationship between voter and his vote. Also, the nodes in a blockchain-based voting system should be connected to each other being susceptible to cyber security issues and threats.

In general, the use of blockchain in public services and governmental affairs offers many benefits. From a technical perspective, it enables simple and transparent methods to efficiently audit public services, and also as an interoperability platform. The main benefits of blockchains in government services are:

- *Transparency*: it is a key requirement for democratic governments and also a fundamental characteristic of the blockchain. Every data stored in the network is replicated to all nodes that take part of it, making the data or a hash (i.e., a pointer linking to the data as a proof of existence) of it, being transparent to the public.
- *Interoperability*: once different governmental systems maintain a single view on given information (e.g., financial expenses report of deputies), it can be automatically accounted on different systems without duplicates.
- *Auditability*: once data is stored, it provides a permanent record that serves as a constant auditable trail to the population.

However, there are many challenges to be solved. While transparency is desired for many public services (e.g., public expenses), it is still needed to maintain privacy in services provided to citizens (e.g., health care, identities, vote). While blockchain technology can preserve the integrity and confidentiality of records using cryptographic methods, as shown in the Estonian use cases, it is also necessary to understand the security guarantees provided by the overall system, which include the integration between blockchain and other centralized systems. Also, to fully take advantage of all blockchains benefits, it is not only necessary to understand the deployment model (i.e., public or private permissioned) verifying which entities and how decentralized are the nodes responsible for the network, but also the security of the centralized bodies which insert data in the blockchain. Although there is still the possibility of manipulation or errors on the input of the data, it is important to highlight that technology is, in general, an excellent tool for improving the transparency and interoperability of public services.

2.2.6 Computer Science

Many computer science applications can benefit from the decentralization leveraged by blockchains. Besides the IoT-based applications, other cases can be seen in literature taking advantage of the blockchain capabilities. For example, in the context of a distributed denial-of-service (DDoS) defense, blockchain could be used as a platform to exchange mitigation services in a cooperative network defense (i.e., service providers cooperating toward a global network defense). A cooperative network defense has many benefits, for instance, by utilizing other organization's resources the burden of the protection can be shared, and defense capabilities can be extended through the different protection systems participating in the distributed defense. Thus, blockchain would serve as an immutable platform for the exchange of mitigation services expressed in SCs.

The Blockchain Signaling System (BloSS) [4] is one example that leverages blockchain capabilities to define individual agreements or conditions in a SC to perform a mitigation service in a cooperative network defense. Conditions service are described in an individual SC which expresses the networks managed by a peer and the conditions (e.g., cost to operate the infrastructure) to perform a mitigation service. In a mitigation request, service and validation time-outs are agreed between the peers to establish deadlines for mitigation service completion and the payout of the reward (cost) upon the submission of a mitigation proof.

In BloSS, SCs and decentralized applications are the main components. While SCs describe how information is exchanged among autonomous system (AS), the applications contain the parameters that define how an AS interacts with a network management system. An AS creates a blockchain account (wallet) and notifies a central contract that holds information on IP networks being operated by each AS, the address of their wallet, and the address of their contract. Operating IP networks and the token price could be modified in the central contract. Individual SCs implement a method to retrieve addresses reported by others ASes and to be aware of addresses of other domains, ASes can query the central SC to maintain a local lookup table. With the knowledge of IP networks maintained and, possibly, the token prices of other participants, ASes may request protection submitting a transaction to their SC with a list of IP addresses. Subsequently, requested ASes may accept or deny requests based on their security policies or service level agreements.

In this case, blockchain becomes a useful tool not only to propagate attack information, replacing existing gossip-based protocols but also a tool that allows the exchange of information (e.g., cryptocurrencies) for the enforcement of mitigation services. However, there may be restrictions on the confidentiality of the disclosed data since attacks information can have an economic impact on service providers. This makes it necessary to have a permissioned blockchain, restricting permissions to read and write through a consortium responsible for managing the cooperative defense.

Blockchain can be used as a means of protecting many Internet services typically provided in centralized systems, such as DNS (domain name system). Traditional DNS is vulnerable to a variety of attacks due to protocol flaws and the centralized nature, which is usually target of DDoS attacks. For example, DNS records can be distributed across the blockchain network avoiding the single point of failure and increasing the difficulty of potential attackers to make it unavailable. For example, the identity service Blockstack [26] uses a naming system (termed blockchain name system) operating across different blockchains. As it is built on top of the Bitcoin [1], which cannot store large data such as name-value pair information, and separate logical layer (i.e., virtual chain) is used to organize the naming system regardless of the underlying blockchain. Therefore, not only can Blockstack increase the data storage capacity considerably but also allows the logical layer to improve and extend independently.

Essentially, blockchain technology is a candidate of great promise for the next-generation DNS system in that it inherently possesses the qualities of censorship-resistance, security, and resilience. In summary, blockchain-based DNS systems own the required properties for a naming system. They are immune to attacks traditional DNS usually suffers such as protocol attacks and server attacks. Additionally, they can offer privacy protection for the clients as the DNS lookup can be done locally at the chain stored in the nodes. However, while maintaining a complete local copy of DNS entries might impact the performance of DNS lookups positively, it can be costly regarding disk space. For example, Namecoin [30], the blockchain size currently is 5.29 GB, and it is expected that the volume will significantly increase when scaling to millions of users in the future.

Other computer science technologies can also be combined with the blockchain, such as artificial intelligence. For example, AI can be defined in a simple way as a layer on top of blockchain data to analyze data (e.g., looking for irregularities in public services[c]), or to interconnect AIs in a marketplace where different AI can communicate to each other, as described in the project SingularityNET [31]. The goal is to provide an open platform based on blockchain where any AI developer can include modules with specific capabilities, combining all the different knowledge to create a better AI. Therefore, SingularityNET aims to become a standard protocol for networking AI and machine learning tools [31].

Another interesting project is presented in the ScienceMatters project [32]. Currently, scientists must be trusted to provide a true and useful representation of their research results in their final publication; blockchain would make much larger parts of the research cycle open to scientific self-correction. While the current centralized solution solves important issues in academic publishing, a blockchain-based solution can go even further.

Full transparency, where the researcher can see also previous reviews of reviewers, any potential social bias can become visible. Not only the reviewers can be rewarded as in the centralized system (e.g., reward scientists who conduct a timely peer review), also a scheme where a reference to other reviewers can be rewarded. Thus, a ground breaking and often cited paper would be rewarded with more Ethers or Tokens. Trying to achieve the same functionality with the current banking system is much more difficult and error-prone. The same way an author gets Tokens or Ethers for getting cited, this could also be applied for confirmation or contradiction of existing research. For a confirmation, Tokens or Ethers can be added, while for a contradiction Tokens are Ethers may be deducted.

---

[c] A.I. for social control of public administration: https://serenata.ai/en/.

Middlemen are no longer needed to decide what is published and what is not. Instead, as long as observations pass the required peer review process they are published and recorded on the blockchain. This ensures scientists' and researchers' discoveries and observations are recorded in a tamper proof way with a time stamp. This has important ramifications for protecting intellectual property. While the blockchain-based solution also solves the same issues as the current ScienceMatters platform: Overreliance on Trust and Lack of Objectivity, Fear of Getting Scooped, Many Scientific Observations Are Not Published, Reluctance To Publish Reproduced Results, the most important argument for using a blockchain is "profits and control." With a blockchain-based system, the traditional publishing middleman will become obsolete.

Among the diverse areas of computer science, blockchain can operate as a large interoperability platform enabling systems that, usually operate in isolation or with limited interoperability, to exchange information without intermediaries in a transparent way. Also, the ability to decentralize information enable services, once centrally provided as a single point of failure, to become more resilient to attacks. Therefore, it is general benefits of blockchains in computer science applications can be summarized (but not limited to) as:

- *Interoperability*: enabling a transparent platform for individual systems to exchange information verifiably. It not only enables new possibilities for collaborative systems but also helps to transform the way existing systems communicate.

- *Decentralization*: transference of information is securely done without intermediaries on a peer-to-peer basis. Therefore, services that were centrally provided can be fully decentralized by extending resilience to denial of service attacks.

However, there are still challenges related with the performance which can reduce or delay the interactivity of cooperative systems. Also, maintaining a local copy of the chain results in a disk space requirement that can restrict use on constrained devices. For example, while it is beneficial to perform a local lookup on the DNS entries, it is also to store the entire chain locally. Therefore, storing and maintaining large blockchain data at every node might not be viable. Moreover, it is necessary to understand the blockchain traffic (i.e., protocol execution, block-generation period, and size) to estimate the computational overhead in the targeted system. In this regard, challenges can be summarized as:

- *Performance and storage requirements*: for some applications latency and communication overhead can become a deciding factor. In this sense, the evaluation of the blockchain type as well as its consensus algorithm becomes crucial in the decision. As the number of participants and transactions increases, the blockchain also increases, making it costly to store it, which can be a limiting factor for constrained nodes.

- *Privacy and confidentiality*: ensuring privacy and confidentiality remains a challenge. Even being possible to hide information on a blockchain, using well-established cryptographic methods, and instances may spread data sensitive and reveal the behavior and preferences of their owners. Therefore, complete anonymity is not possible, but only pseudoanonymity, evolving toward solutions that reduce the possibility of identifying systems from the identification of patterns in the behavior of the data.

2.2.7 Discussion

As the public image of blockchains is slowly disassociated from a cryptocurrencies, especially a Bitcoin view, there is a strong trend to discover new application areas and use cases, where a blockchain can be as disruptive as in the financial sector. Indeed, there are several use cases in which blockchains can reshape the way how entities can interact and trade, providing a trustworthy platform to transact without relying on a third party. However, the decision on whether the application of a blockchain makes sense for new use cases is still not trivial. It requires not only to check if a particular use case fulfills all key blockchain properties, but it also requires an analysis on whether a particular blockchain implementation satisfies such a use case's requirements.

The replacement of an existing ecosystem, consisting of a set of entities, processes, and legacy systems by another logic based on blockchains, requires a large capacity of articulation between these entities. In other words, the use of blockchains needs to create a measurable advantage that could justify the replacement of a working ecosystem by a new one based on blockchains. Otherwise, it is unlikely that the new technology will even be tested. Still, not even a positive disruption can result in the adoption of the technology, once government actors can introduce new laws to monitor and regulate the use of a certain technology or system.

The selected application areas and use cases not only confirm that blockchains can, in general, increase the efficiency of how entities interact, but it also reveals specific technical, institutional, and human challenges that need to be addressed before a successful deployment. While almost all

technical challenges are related to trade-offs between application requirements (e.g., performance vs confidentiality) and what the blockchain can
deliver regarding its functionality, institutional and human challenges,
respectively, are related to legal and compliance requirements and the lack
of knowledge about the technology itself. Lastly, there is the economic
aspect regarding the processes of adoption and maintenance of the technology that must also consider the impact in the other aspects.

2.2.7.1 Technological Aspects

Different types of blockchains (permissionless, public/private permissioned)
exist, whereas each one is suitable for a scenario. While, on one hand,
permissionless blockchains present by design a high level of transparency
and redundancy, on the other hand, they show limitations regarding performance and confidentiality requirements. There exist other implementations,
where write and reading permissions can be distributed only to those who take
part in the system. The challenge is to maintain these blockchain properties
of transparency and redundancy, while performance and confidentiality
are increased, which often requires a certain level of trust in a centralized
entity managing the blockchain. However, the most important technical
challenge for blockchains is scalability, which is not a goal to be achieved
simply, but considered today a moving target. For example, the main goal of
cryptocurrencies is to achieve the same level of transactions per second as a
centralized system can do, while key blockchain properties are maintained.
Therefore, the difficulty encountered in the system is intrinsically related to
preserving such qualities, in addition to not impair its appreciation as a
currency.

2.2.7.2 Institutional Aspects

Most governments have not defined rules regulating cryptocurrencies or
other blockchain applications or these rules may be conflicting across state
lines, creating a suitable or limiting environment for innovative blockchain-
based start-ups. On one hand, start-ups are free to innovate, but on the
other hand, more established companies hesitate to adopt the technology
until they know exactly how it will be governed and especially how much
risk will be involved. For example, if an error occurs in the execution
of a SC in a permissionless blockchain, it is not easy to determine who will
be legally responsible. However, government agencies and international
organizations already started preliminary work to regulate the technology

(mostly with respect to cryptocurrencies). For example, Japan recognizes Bitcoin as a legal currency being definitely legalized in 2017 [33], and the World Economic Forum (WEF) published a long study on the likely impact of various methods of blockchain governance [34]. Also, changes in regulation can create opportunities, as seen in the cases of modum.io [2] and VALID [24]. Both companies explored changes on European Union regulations to control the distribution of medicinal products [2] and ownership of digital user data [24].

2.2.7.3 Human Aspects

From a practical point of view, many people and places are not ready for a digital transformation. As described in the supply chain area [2], most of the personnel working in warehouses are very well used to paper-based processes, and replacing these processes by digital ones (e.g., an app in a smartphone to create shipments) may require lots of effort on the user experience of these digital processes. Also, most warehouses do not provide network connectivity, which impacts on the hardware design's side of sensors the use of long-range protocols (e.g., LoRaWAN). Also, it is important to note that a blockchain is only responsible for ensuring the information, but the accuracy of the information relies on the input of data, e.g., the sensors applied. Other examples are found within the financial sector, where the user experiences the secure key management for cryptocurrencies. Instead of focusing only in the functionality, such software solutions should also focus on user experience to ease the operation of these features by non-technical persons.

2.2.7.4 Economic Aspects

Blockchain can facilitate, simplify, and improve business and processes, be applying to several areas and cases where multiple entities are involved. Thus, not only the costs associated with manual paperwork, legal documentation processing, duplicate entries, and mismatches can be reduced with digital processes, but it also enables a transparent platform for centralized systems to exchange information verifiably, which also reduces costs. However, it is necessary to evaluate impacts of the blockchain adoption process on the mentioned areas (technological, human, and legal) and their associated costs before beginning the transition process. Also, blockchain is still a technology in the process of evolution and maturation, which may, therefore, require time for the totality or even a portion of its benefits to be fully obtained.

3. RESEARCH DIRECTIONS

While the majority of blockchain research focuses on revealing and improving technical limitations of blockchains, such as *scalability*, *privacy*, and *security* requirements, other aspects are becoming equally important. Among these technical requirements, scalability is the main concern of both industry and academia, seeking solutions to increase the number of transactions per second (especially for a public blockchain), while the consensus over the chain and current blockchain properties are to be maintained unchanged. In addition, other research directions, which may pave the way for more use cases being applicable, are not to be forgotten, thus, they are discussed here.

With the rise of attention in the blockchain technology and the lack of regulations in many sectors, many blockchain-based start-ups are using their own blockchain platform to raise funds through an initial coin offering (ICO). Also, different blockchains for different application areas and use cases will create the need for a blockchain interoperability and new processes of distributed governance to organize the interactions of many entities using many blockchains.

3.1 Initial Coins Offering

Among the various potential applications of a blockchain, one of the most promising ones and a very frequently adopted one is an ICO—initial offers of currencies (similar to traditional initial public offerings—IPOs). ICOs can be considered as a new way to raise funds to finance project development. So far, ICOs were used primarily by small businesses and start-ups for creating and issuing a token through the blockchain for an initial fundraising campaign. Sometimes, the new token is offered to the public in exchange of existing cryptocoins (e.g., Ether or Bitcoin), but this is not always the case. Tokens issued in ICOs may also offer rights in a company, i.e., voting and capital rights, or even access to different services and applications, which could be enabled through SCs.

Cryptocurrencies and ICOs are increasingly subject to inspection by public agencies. They are starting to publish their interpretations and rules to be applied from a governmental regulation perspective. In Switzerland, for example, even if ICOs are not officially regulated by Swiss legislation, the independent financial markets regulator (FINMA) published guidelines defining how it intends to apply financial market legislation in handling inquiries from ICO organizers [35]. The goal of this document is to create transparency in how an ICO is conducted, protecting both the organizers and investors from fraudulent activities (such as money laundering).

3.2 Blockchain Interoperability

In the future, assets and information are predicted to be spread in different blockchains of different companies (e.g., financial information or digital information), revolving a similar problem of current traditional databases. In this regard, the blockchain interoperability or federation has the goal to collect data from multiple parallel blockchains, while retaining each of their security properties. Thus, it means that users could access a wide range of blockchain features that are native to each chain without the need to download large blockchain files for every chain they want to use [36]. Currently, interoperability can be handled off-chain with developers handling different SC languages or third-party tools, but it requires a standardization effort toward a common SC language across all blockchains. In addition, the handling of information off-chain can create inconsistencies in case of network failures or longer-term lacks of access.

3.3 Blockchain Governance

For the first time in history, communities can now reach consensus and coordination at a global level through cryptographically verified peer-to-peer procedures to provide governance services in a more efficient and decentralized way, without the intermediation of a third party. However, blockchains still require a set of common rules by which all participants operate, to ensure its accuracy and trustworthiness, and it is not yet clear what and how the governance process is going to be in the face of the many possible application areas and use cases. For example, despite of interoperability efforts promoted in the health care industry by the Office of the National Coordinator for Health Information Technology (ONC) and the blockchain technology being able to satisfy most requirements, it is still not clear how the governance process can be defined to organize how many entities can cooperate in such complex industry.

4. CONCLUSIONS

In overall conclusion, current work on blockchains and SCs does indicate that many fully decentralized actions can be performed reliably in a distributed manner, however, efficiency, security, and legality in a multitude of different facets—typically determined by the dedicated application area or use case—have to be addressed in an integrated manner in the near future to reach a long-term viable balance of usability and efficiency of major or even all blockchain operations. Thus, the roles of the various use cases discussed earlier, the technological alternatives of blockchains being implemented and used, and technical, regulatory, human, and economic requirements require a thorough evaluation to determine whether a blockchain solution makes sense or not for a particular use case.

REFERENCES

[1] N. Satoshi, Bitcoin: A Peer-to-peer Electronic Cash System.: 28, URL: https://bitcoin.org/bitcoin.pdf, 2008.

[2] T. Bocek, B. Rodrigues, T. Strasser, B. Stiller, in: Blockchains Everywhere—A Use-Case of Blockchains in the Pharma Supply-Chain, IFIP/IEEE Symposium on Integrated Network and Service Management (IM 2017), 2017, pp. 772–777. Lisbon, Portugal.

[3] W. Karl, A. Gervais, Do You Need a Blockchain? IACR Cryptology ePrint Archive: 375, 2017. Cryptology ePrint Archive.

[4] G. Wood, Ethereum: A Secure Decentralised Generalised Transaction Ledger, Ethereum Project Yellow Paper 151 (2014) 1–32. Chicago.

[5] T. Bocek, B. Stiller, Smart contracts—blockchains in the wings, in: Digital Markets Unleashed, Springer, Heidelberg, Germany, January, 2017, pp. 1–16.

[6] Z. Zheng, S. Xie, H.N. Dai, H. Wang, Blockchain Challenges and Opportunities: A Survey, 2016. Working Paper.

[7] D. Drescher, Blockchain Basics: A Non-Technical Introduction in 25 Steps, Apress, New York, NY, 2017.

[8] G. Greenspan, Avoiding the Pointless Blockchain Project, URL: https://www.multichain.com/blog/2015/11/avoiding-pointless-blockchain-project/, 2015.

[9] B. Vitalik, On Public and Private Blockchains, URL: https://blog.ethereum.org/2015/08/07/on-public-and-private-blockchains/, 2015.

[10] P. Boucher, How Blockchain Technology Could Change Our Lives. European Parliamentary Research Service (EPRS), Scientific Foresight Unit (STOA), 2017. PE 581.948. URL: http://www.europarl.europa.eu/RegData/etudes/IDAN/2017/581948/EPRS_IDA(2017)581948_EN.pdf.

[11] M. Iansiti, K.R. Lakhani, The truth about Blockchain, Harv. Bus. Rev. 95 (1) (2017) 118–127.

[12] K. Korpela, J. Hallikas, T. Dahlberg, in: Digital Supply Chain Transformation Toward Blockchain Integration, 50th Hawaii International Conference on System Sciences, Hawaii, U.S.A., 2017.

[13] IBM, Maersk and IBM Unveil First Industry-Wide Cross-Border Supply Chain Solution on Blockchain, URL: https://www-03.ibm.com/press/us/en/pressrelease/51712.wss, 2017.

[14] S. Dani, Food Supply Chain Management and Logistics: From Farm to Fork, Kogan Page Publishers, 2015. URL: https://www.koganpage.com/media/project_kp/document/food-supply-chain-management-and-logistics-sample-chapter.pdf.

[15] M. Schneider, Design and Prototypical Implementation of a Blockchain-Based System for the Agriculture Sector, Universität Zürich, Communication Systems Group, Department of Informatics, Zürich, Switzerland, 2017.

[16] European Union, Guidelines of 5 November 2013 on Goods Distribution Practice of Medicinal Products for Human Use, URL: http://ec.europa.eu/health/files/eudralex/vol-1/2013c343 01/2013c34301en.pdf, 2013.

[17] K.B. DeSalvo, RE: Connecting Health and Care for the Nation: A Shared Nationwide Interoperability Roadmap, 2015. URL: https://www.healthit.gov/sites/default/files/hie-interoperability/nationwide-interoperability-roadmap-final-version-1.0.pdf.

[18] S. Angraal, M.H. Krumholz, W.L. Schulz, Blockchain technology: applications in health care, Circ. Cardiovasc. Qual. Outcomes 10 (9) (2017) e003800.

[19] A. Ekblaw, A. Azaria, J.D. Halamka, A. Lippman, A case study for blockchain in healthcare: "MedRec" prototype for electronic health records and medical research data, in: IEEE Open & Big Data Conference, 13, 2016, p. 13.

[20] M. Gault, Increasing Healthcare Security With Blockchain Technology, URL: https://guardtime.com/blog/increasing-healthcare-security-with-blockchain-technology, 2017.

[21] A. Dorri, S.S. Kanhere, R. Jurdak, Blockchain in Internet of Things: Challenges and Solutions, 2016. arXiv preprint arXiv:1608.05187.

[22] K. Christidis, M. Devetsikiotis, Blockchains and smart contracts for the internet of things, IEEE Access 4 (2016) 2292–2303.

[23] C. Lundkvist, R. Heck, J. Torstensson, Z. Mitton, M. Sena, Uport: A Platform for Self-Sovereign Identity, Whitepaper. URL: https://whitepaper.uport.me/uPort_whitepaper_DRAFT20170221.pdf, 2017.

[24] VALID, Your Data Your Asset, Whitepaper. URL: https://valid.global/static/valid-wp-2.pdf, 2018.

[25] B. Hall, 5 Identity Problems That Blockchain Does Not Solve, URL: https://blog.id.me/identity/fintech/5-identity-problems-blockchain-doesnt-solve/, 2017.

[26] M. Ali, J. Nelson, R. Shea, M.J. Freedman, Blockstack: A Global Naming and Storage System Secured by Blockchains, USENIX Annual Technical Conference, Denver, CO, 2016.

[27] E-Resident, Republic of Estonia—e-Residency, URL: https://e-resident.gov.ee/, 2018.

[28] I. Martinovic, Blockchains: Design Principles, Applications, and Case Studies, Working Paper No. 7 (Draft). URL: http://egov.ee/media/1374/martinovic-blockchains-design-principles-applications-and-case-studies.pdf, 2017.

[29] Y. Liu, Q. Wang, An E-Voting Protocol Based on Blockchain, URL: https://eprint.iacr.org/2017/1043.pdf, 2017.

[30] A. Loibl, J. Naab, Namecoin. Seminar Innovative Internettechnologien und Mobilkommunikation SS2014, URL: https://www.net.in.tum.de/fileadmin/TUM/NET/NET-2014-08-1/NET-2014-08-1_14.pdf, 2014.

[31] B. Goertzel, S. Giacomelli, D. Hanson, C. Pennachin, M. Argentieri, SingularityNET: A Decentralized, Open Market and Inter-Network for AIs, URL: https://public.singularitynet.io/whitepaper.pdf, 2017.

[32] Science Matters, The Next-Generation Science Publishing Platform, URL: https://www.sciencematters.io/, 2018.

[33] B. Yap, New EU Payment Services Rules Spur New Regulation in Japan, International Financial Law Review, 2017. 3/6/2017, p. 7. URL: https://searchworks.stanford.edu/articles/bah__121561457.

[34] World Economic Forum (WEF), Realizing the Potential of Blockchain. A Multistakeholder Approach to the Stewardship of Blockchain and Cryptocurrencies, Whitepaper, 2017. URL: http://www3.weforum.org/docs/WEF_Realizing_Potential_Blockchain.pdf.

[35] FINMA, Guidelines for Enquiries Regarding the Regulatory Framework for Initial Coin Offerings (ICOs), URL: https://www.finma.ch/en/news/2018/02/20180216-mm-ico-wegleitung/, 2018.

[36] V. Buterin, Chain Interoperability, URL: http://www.r3cev.com/s/Chain-Interoperability-8g6f.pdf, 2016.

FURTHER READING

[37] B.B. Rodrigues, T. Bocek, B. Stiller, In: Blockchain Signaling System (BloSS): Enabling a Cooperative and Multi-Domain DDoS Defense, 42nd Annual IEEE Conference on Local Computer Networks (LCN 2017, Demo Track). Singapore, 2017.

ABOUT THE AUTHORS

Bruno Rodrigues: received the B.Sc. degree in computer science from University of the State of Santa Catarina (UDESC) in 2013 and his M.Sc. degree in Computer Engineering from the Polytechnic School of University of São Paulo in 2016. He joined the Communication Systems Group (CSG) at the University of Zurich (UZH) in September 2016 to obtain his Ph.D. degree.

Prof. Dr. Thomas Bocek: received the Doctor of Informatics (Dr. Inform.) degree from the University of Zurich. Dr. Thomas Bocek is currently the head of P2P and distributed systems at the Communication Systems Group at the University of Zurich since 2013. Before that, he worked as a software engineer and technical project manager in the financial sector. Thomas is mainly interested in communication systems and networks, especially focusing on peer-to-peer, distributed systems, and blockchains including Bitcoin and Ethereum. He is also involved in the blockchain start-up modum.io, which combines IoT sensor devices with blockchains. In 2018, Thomas will join the University of Applied Sciences of Eastern Switzerland as a professor for distributed systems.

 **Prof. Dr. Burkhard Stiller**: received the Diplom-Informatiker (M.Sc.) degree in computer science and the Dr. rer.-nat. (PhD) degree from the University of Karlsruhe, Germany. He chairs as a full professor the Communication Systems Group CSG, Department of Informatics IfI, University of Zurich UZH since 2004. He held previous research positions with the Computer Laboratory, University of Cambridge, UK, the Computer Engineering and Networks Laboratory, ETH Zurich, Switzerland, and the University of Federal Armed Forces, Munich, Germany. He did coordinate various Swiss and European industrial and research projects, such as AAMAIS, DAMMO, SmoothIT, SmartenIT, SESERV, and Econ@Tel, besides participating in others, such as M3I, Akogrimo, ECGIN, EMANICS, FLAMINGO, and ACROSS. His main research interests are published in over 200 research papers and include charging and accounting of Internet services, systems with a fully decentralized control (blockchains, clouds), network and service management, economic management, telecommunication economics, and IoT.

CHAPTER SEVEN

Security and Privacy of Blockchain and Quantum Computation

Kazuki Ikeda

Department of Physics, Osaka University, Toyonaka, Osaka, Japan

Contents

Abstract

Security and privacy are vital to the modern blockchain technology since it can exist without an authorized third party, which means that there may not be a trusted responsible person or organization in charge of systems. In this chapter we make a survey on this issue about blockchain systems. Security of the current systems is based on the computational hardness assumptions and many of the standard cryptography systems are known to be vulnerable against the advent of full-fledged quantum computers. On the other hand, it is possible to make a blockchain more secure by virtue of quantum information technology. In this chapter we give a pedagogical introduction to quantum information theory and quantum computation so that readers can follow advanced

Advances in Computers, Volume 111
ISSN 0065-2458
https://doi.org/10.1016/bs.adcom.2018.03.003

researches on application of quantum technology to the blockchain industry. We also explain a novel information system which accommodates quantum states in a peer-to-peer way. It would improve the level of privacy and security by the laws of physics, which is never achievable from nonquantum information theoretic viewpoints.

1. INTRODUCTION

A blockchain is a system which allows users to form a public consensus without going through a particular third party such as a government and a company was invented by the anonymous author in 2008 for the purpose of making the decentralized cryptocurrency, called Bitcoin. Nowadays it offers a wide variety of uses and many similar systems have been invented; virtual currencies are widely used for multiple payments and products can be traded or sheared without brokers. Moreover some countries have considered introducing the blockchain technology to public services. Indeed, Estonia issues e-Residency, a digital ID that offers people all over the world to start up a location-independent business. The citizens can use it for various procedures including election, tax payment, bank transfer, and so forth. More recently the government of Venezuela has officially launched Petro, a new and the world's first government issued digital currency. As just described, a blockchain has a big potential advantage; therefore, it is highly important to ensure security of the systems.

Regarding security and privacy, simplicity of a blockchain's structure helps us to specify its weak points. Namely we should pay very close attention to only the following two points: the digital signature scheme and the proof of work. Those two aspects have different connotations in terms of security and privacy. First of all, security of the digital signature scheme used for blockchains is guaranteed by computational hardness of elliptic curve cryptosystems. If our world is by no means immune from quantum computing, this classical digital signature is secure since there are no supercomputers designed for good performance in solving such NP-hard problems. Moreover it has been believed that breaking elliptic cryptography would be beyond the capability of the mid-21th century quantum technology. However, advantages of quantum information theoretic viewpoints can be found at many different aspects, one of which is privacy. In the conventional blockchain technology, there are no ways to protect users' privacy perfectly (indeed this problem was pointed out in the original Bitcoin paper and is still an open problem). Hence the system experiences vulnerability. However, quantum technology does ensure privacy and solves safety problems by the laws of physics. For example, a quantum key distribution (QKD) protocol allows a remitter and a receiver to shear information safely. Moreover the conventional digital signature protocol based on the elliptic curve should be replaced by a quantum digital signature protocol so that no one but the owner can tamper. In addition to that blind quantum computation protects users' privacy in a perfect way since they can delegate their quantum computation to a server with a quantum computer in such a way that the server cannot recover any input and output of the clients.

2. INTRODUCTION OF SECURITY AND PRIVACY OF BLOCKCHAIN

2.1 Digital Signature and Elliptic Curve Cryptosystems

A blockchain is a relatively new technology which combines traditional methods in a sophisticated way. Therefore the problems on security and privacy of a blockchain are literally true to any conventional systems using the traditional cryptography. However, people's fear of security and privacy risk to a blockchain system seems to be greater than that of a bank or any organization. This would be because that a blockchain can exist without supervision from an authorized third party. Banks or companies protect clients' information by fail-safe plans so that they limit access to information to the trusted third parties, whereas when it comes to a blockchain, robustness of the system is crucial since it does not assume those parties. The main purpose of this section is to give a comprehensive explanation on the security letting a blockchain be a decentralized and distributed database and the associated problems on its privacy. This part doubles as an introduction to blockchain's architecture.

As the first step to discuss our concept more precisely, we shall obtain the minimal knowledge on the elliptic curve cryptography [1,2], which is the underling scheme making the whole system secure and fundamental for digital signature algorithm. As we will discuss later, the security of the blockchain protocol totally depends on the security of the elliptic curve cryptography. We are interested in curves, called elliptic curves, as shown in Fig. 1. Elliptic curves are very rich in mathematical study and play a fundamental role in solving Fermat's last theorem. For more advanced mathematical materials, readers should consult with advanced references [3, 4]. Multiplication of two points A and B in a given elliptic curve is defined as the negation of the point C resulting from the intersection of the curve, and the straight line defined by the points A and B, giving the point D.

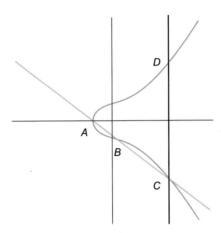

Fig. 1 Elliptic curve.

Namely we have $A+B=D$. We include the infinity point $O=(\infty,\infty)$. So in the above example, $A+A=O$. In general it is known that, for any point P in a curve, there is the minimum number n which satisfies $nP=O$. We call such n as the order of the point. It will be easy to guess that computing $Q=rP$ from given r and P is easy; however, it will be quite hard to generate r if one only knows P and such Q. The elliptic curve cryptography is based on computational hardness of this problem. Now, let E be an elliptic curve of the form defined over a finite field $\mathbb{F}_p = \mathbb{Z}/p\mathbb{Z} = \{0, 1, ..., p-1\}$

$$E: y^2 = x^3 + ax + b,$$

where p is a prime and parameters a and b satisfy $4a^3 + 27b^2 \neq 0 \pmod{p}$. The total number $\#E$ of points of E is called the order of E and the order n of P is a divisor of $\#E$. To generate safer signatures, $\#E$ should be a prime. Suppose Alice wants to deliver a message m to Bob. To conceal the message, Alice gives a signature to her message m and sends it to Bob by the following rules. In order to convey her message successfully, Alice and Bob shear the information of a prime p, parameters $(a, \text{and } b)$, a base point P in the given elliptic curve, and its order n. In blockchain systems, SHA256 hash function $h: \{0,1\} \rightarrow \{0,1\}^{256}$ is commonly used. The prime p should be taken sufficiently large to make the signature secure.

2.1.1 Generate Signature
1. Prepare her private key s $(1 \leq s \leq n-1)$ and compute a public key $R=sP$.
2. Calculate hash values $h(m)$ of the message m.
3. Generate a random integer t $(1 \leq t \leq n-1)$ and compute $tP=(x,y)$.

4. Compute $c \equiv x \bmod n$ and $d = t^{-1}(h(m) + sc) \bmod n$.

5. Alice sends m, her signature (c, d), and the public key R to Bob.

It is important to keep t secret; otherwise the private key s can be guessed easily by $s = c^{-1}(td - h(m)) \bmod n$.

Verification process of her signature is relatively simple. Bob only needs to calculate

$$S = f_1 P + f_2 R = (x_1, y_1),$$

where $f_1 = d^{-1} h(m) \bmod n$ and $f_2 = cd^{-1} \bmod n$. Let $c' = x_1 \bmod p$. Then he approves the signature if and only if $c' = c$.

Now we can clearly understand the basic architecture of a blockchain and discuss the security and privacy issues it faces with. The fundamental architecture of a blockchain is displayed in Fig. 2, and it works by passing signatures and transaction records to other persons in a peer-to-peer way. Namely, a person (Alice) who wants to transmit coin to Bob signs her transaction request and broadcasts it to the public. Then the other participants collect her request data and verify her signature using her public key. If more than 51% of them approve the signature, then her request is accepted. A third person renews the transaction record and adds a new block to the chain, by which Bob can receive the coin from Alice. There are basically two different themes of security and privacy issues. One of them is the possibility that the system can be hijacked by the majority possessing huge computing power. This is an inevitable problem of conventional blockchain systems since they are based on public will. A possible solution is proposed

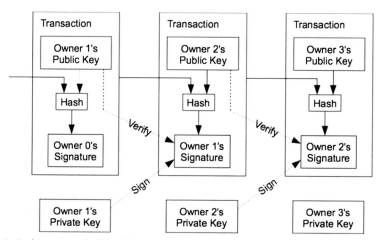

Fig. 2 Architecture of a blockchain [5].

in Chapter "Applications of blockchain in the financial sector and a peer-to-peer global barter web" by Ikeda and Hamid, where a novel blockchain protocol is applied to economic study and the 51% problem is solved in a certain situation. Another problem is hidden in the cryptography a blockchain employs. As shown in Fig. 2, underling scheme is totally based on the elliptic curve cryptography we discussed before. Note that any public key distribution protocol such as RSA, which is supported by computational hardness of prime factorization, is fine as long as it is safe. Indeed most blockchain platforms rely on the elliptic curve cryptography (ECDSA) or RSA. However, the elliptic curvature cryptography is preferred in most of the protocols since it is possible to ensure safety and efficiency with short keys, thereby allowing high-speed processing. In a simple blockchain as shown in the figure, blocks consist of signatures and transaction records, and transactions are completed by authorization of owners' signatures. The fundamental problems of this protocol can be summarized as follows. First of all, as we mentioned several times, the security of a blockchain is directly related to that of the digital signature. Therefore, apart from quantum computing, owners should keep their keys in safe custody. For example, they should keep public keys anonymous in order to maintain privacy. Keeping private keys secret is obviously important. We will come back those points later from a perspective of quantum computing and explain the reasons why ECDSA and RSA are at risk of quantum attack.

2.2 Several Vulnerabilities of Blockchain Systems

Now consider more practical use cases. Apart from theoretical brittleness we mainly discuss in this chapter, there are a lot of weak points hidden in actual blockchain systems. In this subsection, we take a close look at this issue. As shortly described previously, one of the most famous problems is called as the 51% problem. Since each transaction requires approval more than 51% of participants, it is obvious that if a single group with a huge amount of electricity occupies the majority, then it predominates as the cause of the proof of work and unfair transactions can be accepted in a legal way. However, the real situation is worse and the actual threshold becomes lower than 50%. So-called the selfish mining protocol allows a selfish attacker with more than 33% of the total mining power of the network to dominate the market [6]:

1. A selfish miner seeks a flesh block and mines a new block without opening it to the public.
2. Honest miners try to mine the block mined by the selfish miner.

3. While the honest miners do mining, the selfish miner completes the business and repeats the process for many times.

Even though this kind of attacking protocol is not assumed in the original work of Bitcoin, it is reported that implementation of this algorithm is possible in practice. The key point of this strategy is that a selfish participant finds and mines a block before somebody else. Similarly, writing down earlier transaction times (within the allowed intervals) into blocks is a widely used cheat code, by which a block mined by a greedy miner can be recognized as the genuine one by other miners and the greedy miner is likely to win a reward. Miners in the original blockchain are assumed to be completely honest, but the attacking protocols above tell us that those who want to design a blockchain should take a special care in order to maintain the system successfully (Fig. 3).

As a relevant problem, transaction malleability is one of the vulnerabilities and a way to change the transaction hash, by which one can repeat the same transaction in a different way (the double spending). Suppose Bob is about to give his 1BTC which was owned by Alice before to Charlotte. Then the transaction consists of an input and an output (see Fig. 4). The previous transaction ID refers to the transaction ID from Alice to Bob and the index is the reference of the output from Alice to Bob. Regarding to the output, 1BTC should be recorded as the value. The transaction malleability

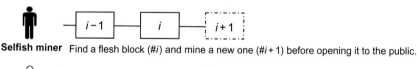

Selfish miner Find a flesh block (#*i*) and mine a new one (#*i* + 1) before opening it to the public.

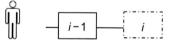

Honest miner Find and try to mine the *i*th block.

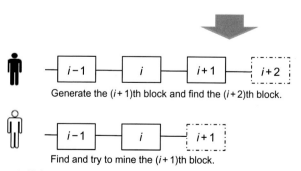

Generate the (*i* + 1)th block and find the (*i* + 2)th block.

Find and try to mine the (*i* + 1)th block.

Fig. 3 Selfish mining.

Input	Output
Previous transaction ID	Value
Index	Public key
Signature	

Fig. 4 Input and output in a transaction. The input (output) refers to the record of the previous (next) owner of the coin.

appears if one changes only the signature. In a naive blockchain architecture, each transaction including the input and the output is hashed in a unique way. Therefore, if one changes the signature, the outcome (hash value) to be used to verify the transaction is changed without any modification to the output. But miners recognize the changed signature valid, and the modified transaction is also approved. This transaction malleability is pointed out as a trigger of Mt. Gox's bankruptcy in 2014 after losing 640,000 bitcoins.

There is a solution to solve the transaction malleability. SegWit (Segregated Witness) was invented to solve it and atomic swaps, which attempt to make transactions between different blockchains possible, are widely surveyed. In 2017, SegWit was activated and it has been attracted a general interest from miners and blockchain developers.

2.3 Conventional Methods for Protecting Privacy

As referred, there are number of problems in the traditional blockchain architecture apart from quantum perspectives we want to discuss next. Before we move on the quantum part, we consider and deepen our knowledge on several efforts to solve those issues. These endeavors are also very important not only for scientific viewpoints but also for realistic implementation and will be helpful when we discuss the matter from a quantum perspective. Roughly speaking, basic privacy problems of many of existing blockchain systems are summarized as the following table.

Remitter	△	Indirect connection to personal information, but possible to guess
Receiver	△	Indirect connection to personal information, but possible to guess
Amount	×	Easy to check on blockchain
Date and time	×	Easy to check on blockchain

Looking at the table, readers would understand that storing data in a blockchain is not a very good way compared with the traditional centralized ways for transactions. For sure, a third party cannot know personal information of remitters and receivers; however, there are still chances to guess from their wallets or addresses. Even worse, amount of transactions and transaction time are easily guessed since transaction records are generically open to public. This is quite different from the usual centralized system. For instance, a bank that makes a special effort to protect information should not open any of clients' information to any bystander in all weather. Howbeit a person who does not trust banks should distribute cash by hand after a considerable lag. Therefore in every face of information protection, a naively designed blockchain would not become the first choice.

Those negative points, however, can become positive if a blockchain is used in a different way. Guidance will be given in succeeding chapters. A good system in general should offer various usages in as many fields as users want. Therefore the next matter we are encouraged to discuss in the rest of this subsection will be how to fix the problems associated with privacy. Issues we mainly consider are listed as follows:

1. How to secret information of owners so that anyone irrelevant to a transaction cannot know who are involved in it.
2. How to secret amount of transactions.

2.3.1 Ring Signature

Sometimes participants in a network want to make their signatures anonymous for several reasons. For example, a successful bidder participating an auction may not prefer to disclose his or her name. A ring signature is a method to satisfy this requirement in such a way that a person who is about to sign a document makes a group and generates a digital signature [7]. Therefore a document signed with a ring signature is created by someone in a group and it is difficult for an outsider to point out its owner. Traditionally a certain central server manages ring signatures and recently some decentralized protocols are proposed.

2.3.2 Zero-Knowledge Proof and Confidential Transaction

It is natural to ask how to make Bitcoin transactions secret. The confidential transaction gives a realistic solution to this question. It is based on the elliptic curve cryptography; hence readers would find value in gaining a better comprehension of both the elliptic cryptography explained previously and the confidential transaction protocol. The essential idea of confidential transaction is based on zero-knowledge proof: if a given statement is true, an honest person can convince the fact without knowing any information apart from the fact that the statement is indeed true. While it ensures anonymity, each transaction needs very long time to wait for transaction approval. Roughly speaking, the zero-knowledge proof in which we are interested here is a protocol to show the formula

$$a + b = c$$

without knowing the exact values of a, b, c. Of course no one can verify it in such a condition, so instead, verifiers show the correctness of the formula by confirming a "similar" formula defined by means of a, b, c. For example, verifiers show $aK + bK = cK$, where K is a certain value, and convince that the equation they have proved may come from the equation $a + b = c$ successfully. Now let us explain how it is possible. As previously, we consider an elliptic curve E over the field $\mathbb{F}_p$. Alice decides a point P of the order n on E for her private key. Let $Q = rP$ be her public key (she needs to keep r secret). Alice also generates a point $H \in E$ using P and a hash function (maybe SHA256) and adds to Q

$$\text{commitment} = rP + aH,$$

where a is the amount of coins to commit. Now Alice uses the value of this commitment for her new public key. We simply write it as $C(r, P, a, H)$. This makes sense since it is hard to guess r, P, a, H from it. Note that $C(r, P, a, H)$ is again a point in E. Then by the additivity of points on elliptic curves $(C(r, P, a, H) + C(r, P, b, H) = C(r, P, a + b, H)$, for example), verification can be completed if verifiers confirm

$$\text{input} - \text{output} = 0.$$

In this way, one can verify that the amount of transaction is correct only by public source. Zcash employs the zero-knowledge proof and protects privacy in a practical manner and several mega banks as represented by J.P. Morgan show interests in using the system for transactions and payments. Though this protocol realizes anonymity in some sense (unless the elliptic curve cryptography is broken), one needs to wait longer than the usual Bitcoin transactions in order for verifiers to approve the correctness of the given statements.

2.3.3 Mixing Coins

Coin mixing is a prevalent method to ensure anonymity for coin owners by randomly shuffling input addresses of owners. The simplest protocol requires a central server to delegate their transactions while mixing: owners tell addresses and coins to the mix, who shuffles information and transfers coins. While there are use cases in practice, they face several problems: this protocol assumes the mix to be trustworthy and not to try to steal coins and uncover owners' information. The anonymity of owners is ensured if the mix keeps showing loyalty. Several developments have been done and CoinJoin [8] is an example. In addition to that, a mixing method available for a decentralized system has been invented recently, and it is called as CoinShuffle [9]. This protocol goes as follows. Suppose each of Alice, Bob, and Charlotte has a single bitcoin. Let A, B, C denote their input addresses and A', B', C' be output addresses. Before making payments, they encode all information (Fig. 5).

As illustrated in Fig. 3, Alice first sends her output address to Bob, who mixes his and her output addresses and passes them to Charlotte. Finally, Charlotte randomly shuffles all of them and broadcasts to the network if Alice and Bob confirm that their output addresses are not missed out. This mixing method surly improves anonymity of transactions, though it does not give a complete solution since if Alice and Bob know each other very well, they can lean Charlotte's output address.

In summary of this section, several standard cryptocurrencies which try to improve security and privacy are listed in the following table, where × means the corresponding protocol fails to ensure anonymity and ○ means it contributes to improving privacy.

Currency	Method	Address	Record	Amount
Dash	CoinJoin	×	○	×
Monero	Ring signature	×	○	×
Zcash	Zero-knowledge proof	○	○	○

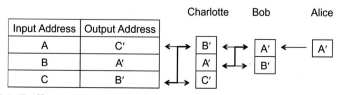

Fig. 5 CoinShuffle.

3. BLOCKCHAIN IN THE QUANTUM ERA

3.1 What Will It Be?

As we move into the third decade of the 21st century, we observe one of the most significant innovations in computer science and industry. That is to say, we are on the verge of becoming possible to acquire full-fledged quantum computers, which are expected to solve some classical problems far faster than any computer in the past. Indeed there are several quantum algorithms as represented by Grover's algorithm [10] which produce a particular output result very faster than classical algorithms. In this sense, progress in quantum computing industry attracts the interest of many researchers in both academic organizations and companies. For example, D-wave quantum computers are good at solving optimization problems by virtue of quantum annealing [11], and it has several practical usages [12–14]. In addition to that there are astounding achievements in the development of quantum communication. There is a method to transport quantum information to a remote area, known as quantum teleportation [15–18]. In future we will broadcast or exchange quantum information on the quantum Internet [19] as we usual do on the World Wide Web [20]. Apart from pure scientific interests, there are several practically important advantages to the quantum world as opposed to the classical one we have been familiar with. One of them is a quantum cryptosystem which is secure by virtue of the laws of physics. We will give more detailed explanations soon. Furthermore a full quantum network linked by quantum channels offers exponentially larger dimension in comparison with a network of quantum nodes linked by classical channels.

As just described, the quantum world seems to have fruitful benefits and we will enjoy productive decades. However, those progresses in the quantum technologies imply that human beings will overcome the classical systems we have inherited from our ancestors and obtain superfast computers which may break most of the classical cryptosystems within a short time; thereby the conventional systems are potentially at risk. According to Ref. [21] the proof of work used by Bitcoin is relatively resistant to substantial speedup by quantum computers in the next 10 years; however, the elliptic curve signature scheme used by Bitcoin is much more at risk and could be completely broken by quantum attack as early as 2027, by the most optimistic estimates. Hence it will be necessary to consider how to protect systems from conceivable quantum attacks. The later part of this chapter is to address this theme. After we gave an introductory review on quantum information, we explain why classical algorithms such as the elliptic curve cryptography and RSA are at a moment of crisis of quantum attack. Last, we describe several possible ways to cure those issue.

3.2 Introduction to Quantum Information

In order to invite readers to a quantum world without fear of getting lost, we begin with describing the fundamental concepts of quantum information, quantum cryptography, and QKD. Throughout this chapter we use quantum states $|\psi\rangle$ taking values in a finite dimensional vector space V. Especially we are mainly interested in a system written by qubits, which are unit vectors in $\mathbb{C}^2$ whose bases are chosen here as

$$|0\rangle = \begin{pmatrix} 1 \\ 0 \end{pmatrix}, \ |1\rangle = \begin{pmatrix} 0 \\ 1 \end{pmatrix}$$

and they correspond to the classical bits 0 and 1. A general qubit $|\phi\rangle$ can be written as a combination of them in such a way that $|\phi\rangle = a|0\rangle + b|1\rangle$, where a and b are complex numbers satisfying $|a|^2 + |b|^2 = 1$. The following two states are often used:

$$|\pm\rangle = \frac{|0\rangle \pm |1\rangle}{\sqrt{2}}.$$

Note that quantum systems have infinitely many ways to generate qubits, whereas classical bits have only two states 0, 1. This is one of the most significant changes from classic information to quantum information. Another important feature is that when quantum states experience measurement, an output result appears with a certain probability. For example, an observer obtains 0 with a probability of $|\langle 0|\phi\rangle|^2 = |a|^2$. A composite state of two qubits $|\psi_1\rangle = (a_1, a_2)^T \in \mathbb{C}^2$ and $|\psi_2\rangle = (b_1, b_2)^T \in \mathbb{C}^2$ is defined by the tensor product

$$|\psi_1\rangle \otimes |\psi_2\rangle := \begin{pmatrix} a_1|\psi_2\rangle \\ a_2|\psi_2\rangle \end{pmatrix} := \begin{pmatrix} a_1 b_1 \\ a_1 b_2 \\ a_2 b_1 \\ a_2 b_2 \end{pmatrix} \in \mathbb{C}^4.$$

We define a general state in a multiqubit system by the multitensor products $|\psi_1\rangle \otimes |\psi_2\rangle \otimes \ldots \otimes |\psi_n\rangle$ and simply write it as $|\psi_1\psi_2\ldots\psi_n\rangle$ or

$|\psi_1\rangle|\psi_2\rangle\ldots|\psi_n\rangle$. Now let us invite a key player, called an entangled state. Using entangled states, two persons A and B can shear their information. We denote V_A by the vector space in which quantum states $|\psi\rangle_A$ of A live. We consider the composite space $V_A \otimes V_B$ of V_A and V_B. By certain unitary operations, a general pure state $|\psi\rangle_{AB}$ in $V_A \otimes V_B$ can be decomposed into a sum of bases

$$|\psi\rangle_{AB} = \sum_{i=0} c_i |\psi_i\rangle_A |\psi_i\rangle_B,$$

here a quantum state is called pure if it is a unit vector. Moreover, we call $|\psi\rangle_{AB}$ an entangled state if more than two of the coefficients c_i are not 0. The following entangled state known as the EPR state [22]

$$|\text{EPR}\rangle = \frac{|0\rangle_A|0\rangle_B + |1\rangle_A|1\rangle_B}{\sqrt{2}}$$

plays a fundamental role when we consider the quantum teleportation later.

Now let us move on to QKD, which is a way to deliver classical bits by means of the quantum nature of photon. The most famous and practical QKD protocol is BB84 invented by Bennett and Brassard in 1984 [23]. BB84 ensures unconditional security of communications by generating keys randomly, which helps to exclude interceptors from communication channels. More precisely, anyone try to capture communications will succeed in interception less than 50% of the time by virtue of the quantum nature (recall that any quantum state is decided with a certain probability) and, even worse for the monitor, people on the formal channel can notice the existence of the interceptor and if they do, they will change channels and continue communications. In practice, quantum states are carried through an optical fiber. There are several developments in the experimental study in QKD. Soon after the announcement of the BB84 protocol, communication between 30 cm at a speed of 3.3 bits/s was succeeded in 1988, and in the 2000s the key rate of about 1 Mbit/s for a distance of 20 km and 10 kbits/s for 100 mk was achieved [24]. More recently, the distribution speed of more than 10 Mbits/s was achieved. In 2017, China successfully sent key data by QKD from space to Earth. Therefore, our communication network has been getting securer year by year and human beings are about to establish QKD as a next-generation communication tool. A standard reference to general quantum information theory is [25], for example (Fig. 6).

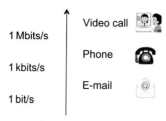

Fig. 6 Required bits/s for practical use cases.

3.3 Why Are the Elliptic Curve Cryptography and RSA at Risk of Quantum Attack?

Now let us consider core reasons why quantum computing has realistic advantages in code breaking. One of the great features of quantum computers is that they have potential advantage in computation by means of superposition of quantum states. As a result, there are indeed some problems which cannot be solved by a classical algorithm in polynomial time on any classical computer, but are solvable by a quantum algorithm in polynomial time on a quantum computer. To get better understanding, we explain several quantum algorithms for code breaking. The central idea is based on the period-finding problem stated below. Basically we consider a function f on $\mathbb{Z}_N = \{0, 1, ..., N-1\}$, $N \in \mathbb{Z}$.

3.3.1 Period-Finding Problem

Let f be a periodic function of period s which is a divisor of N.

$$f(x) = f(x+s), \forall x \in \{0, ..., N-s-1\}.$$

Then determine such period r.

A straightforward classical algorithm for this problem is calculating $f(0)$, $f(1)$, ... and finding s which satisfies $f(0) = f(s)$. This method, however, takes time on the order of $O(\sqrt{N})$ in the worst case. We want to find a faster quantum algorithm.

3.3.2 Quantum Algorithm for Period-Finding Problem

(i) Prepare the following quantum state

$$\frac{1}{\sqrt{N}} \sum_{a \in \mathbb{Z}_N} |a\rangle |0\rangle.$$

(ii) Insert the information of f into the second qubit column

$$\frac{1}{\sqrt{N}} \sum_{a \in \mathbb{Z}_N} |a\rangle |f(a)\rangle.$$

(iii) Suppose one measures an output z. This will give the sate

$$\frac{1}{\sqrt{N/s}} \sum_{a:f(a)=z} |a\rangle |z\rangle.$$

Note that now the summation is taken over all a such that $f(a) = z$.

(iv) Apply the Fourier transformation to the first qubit column and measure it. Then one of the divisors of N/s is obtained.

(v) Repeat (i)–(iv).

As a consequence of this procedure, the values $0, \frac{N}{s}, \ldots, \frac{(s-1)N}{s}$ are obtained with the same probability $1/s$. Suppose one gets an output $m = \frac{kN}{s}$, $k \in \{0, \ldots, s-1\}$. If k and s are coprime, then the problem results in finding the great common divisor d of m and N. s is determined as $s = N/d$. It is known that the Euclidean algorithm is an efficient method to compute the greatest common divisor and it takes time on the order of at most $O(\log N)$. Therefore a quantum algorithm is basically powerful to solve the period-finding problem. One may try to factorize m and N, respectively, to find such d; however, this way is a roundabout approach. The problem of finding the greatest common divisor of two numbers belongs to the class P; however, that of factorization in prime numbers belongs to the class NP.

The great idea to break RSA, which relies on prime factorization, is to convert the problem of prime factorization into that of finding the greatest common divisor.

3.3.3 Factorization Problem

Let N be an integer of n bit length and suppose N can be decomposed into $N = pq$, where p and q are prime numbers. Then determine such p and q.

We are ready to solve this problem by means of the quantum algorithm for the period-finding problem. The trick is to use the function of the form $f(s) = x^s \mod N$. This f has a certain period s. As a simple example, take $N = 15 = 3 \cdot 5$ and $x = 2$. Then it satisfies $f(0) = f(4) = f(8) = \cdots = 1$ and its period is 4. We express the greatest common devisor of two numbers a, b as $\text{GCD}(a, b)$. The following algorithm invented by Shor [26] employs f and solves prime factorization in polynomial time on a quantum computer.

3.3.4 Shor's Algorithm

(i) Pick up a random value $x \in \{1, \ldots, N-1\}$.

(ii) We assume $GCD(x, N) = 1$ since otherwise $GCD(x, N)$ is a nontrivial divisor of N and we can complete prime factorization of N by the Euclidean algorithm.

(iii) Find the minimum s such that $x^s = 1 \mod N$.

(iv) If s is even and $x^{s/2} \neq -1 \mod N$, check whether $GCD(x^{s/2}+1, N)$ or $GCD(x^{s/2}-1, N)$ divides N. If so, output them otherwise start over from (i).

The reason to make a survey on $GCD(x^{s/2}+1, N)$ or $GCD(x^{s/2}-1, N)$ is simple. If s is even, then $(x^{s/2}+1)(x^{s/2}-1) = 0 \mod N$ follows by the assumption (iii). Due to the condition $x^{s/2} \neq -1 \mod N$, $GCD(x^{s/2}+1, N)$ or $GCD(x^{s/2}-1, N)$ corresponds to a nontrivial divisor of N. It is known that one can obtain such s with more than 50% of the time; therefore, Shore's algorithm does make sense and is effective to solve the factorization problem. In this way, we eventually obtained an algorithm to break RSA by means of a quantum computer. Readers should consult with the authors of Ref. [27] for experimental realization of the algorithm.

Now let us consider the elliptic curve cryptography (ECDSA). Basically, ECDSA and RSA have almost the same security level; hence needless to say they are vulnerable against quantum attack as we discussed earlier. But it will be meaningful to describe a quantum method to break ECDSA. The central idea again returns to the period-finding algorithm. Let E be an elliptic curve and P be a point on E. For a given point $Q = rP$, we want to determine such r. Let n denote the order of P and chose N such that $n \leq N \leq 2n$.

3.3.5 Algorithm for ECDSA

(i) Generate the state

$$|\psi\rangle = \frac{1}{n} \sum_{a=0}^{n-1} \sum_{b=0}^{n-1} |a\rangle |b\rangle |aP - bQ\rangle.$$

(ii) Operate the Fourier transformation F to the first and the second columns

$$F|\psi\rangle = \frac{1}{Nn} \sum_{a=0}^{n-1} \sum_{b=0}^{n-1} \sum_{c=0}^{N-1} \sum_{d=0}^{N-1} \exp\left(\frac{2\pi i}{N}(ac + bd)\right) |c\rangle |d\rangle |aP - bQ\rangle.$$

(iii) Measure the states $|c\rangle$ and $|d\rangle$. Then the probability to observe the state $|c, d, R=sP\rangle$ is given by

$$\left| \frac{1}{Nn} \sum_{\substack{a,b \\ a-rb=s}} \exp\left(\frac{2\pi i}{N}(ac+bd) \right) \right|^2$$

(iv) Deduce candidates by using the condition

$$\left| \frac{d}{n} + r\left(\frac{cN - \{cN\}_n}{Nn} \right) \right| \leq \frac{1}{2n}.$$

(v) Candidates of r are suggested by approximating d/n to the nearest multiple of $1/N$. For instance, if $(c, d) = (2, 50)$ and $n = 60$, $N = 31$, then we take $26/31 \sim 50/60$. And divide the result by

$$c' = \frac{cN - \{cN\}_n}{n},$$

where $\{cN\}_n = cN \bmod n$. Using the same example, $c' = 1$. Then r will be

$$r = \frac{26}{31} \times 1^{-1} = -5 \bmod 31.$$

(vi) Accept the triad (c, d, r) if it satisfies the following conditions:

$$\{cN\}_n \leq \frac{n}{12}$$

and

$$\left| T - n\left[\frac{T}{n} \right] \right| \leq \frac{1}{2},$$

where $T = rc + d - \frac{r}{N}\{cN\}_n$. Note that in the example above, $r = -5$ does not satisfy the very last condition; however, if neglect the sign, then $r = 5$, $(c, d) = (2, 50)$, $n = 60$, and $N = 31$ satisfy the last two conditions. To become confident with the result, repeat the calculations for different pairs of (c, d).

After long journey, we eventually and clearly understand several theoretical ways to break RSA and ECDSA. Several experiment data suggest that for a while we may be able to keep having relatively optimistic eyes on defending classical cryptosystems against quantum attack; however, as described we human beings will soon or later acquire complete computing skills to smash them. Therefore we have very good reasons to invent an alternative peer-to-peer system.

3.4 Secured Classical Blockchain From Quantum Attack

In this and the next subsections, we will explain two different ways to shear information in a decentralized and distributed database. We first explain a blockchain which accommodates classical bits and ensures its security by means of QKD. As explained in the previous subsection, the QKD is a way to shear secret random numbers with two users at remote places based on quantum measurement and error correction. One of the most famous protocols was invented by Bennett and Brassard in 1984 and known as the BB84 protocol. It is theoretically proved that, by virtue of the laws of physics, the BB84 protocol allows the two users to share keys with assurance that their information remains secret to anyone else. This fact is in contrast to classical key distribution schemes, which completely rely on the computational hardness assumption and hence it is at risk of being attacked by advance computers. Therefore in order to apply the BB84 protocol to a blockchain and make it secure, one should replace the signature scheme by the QKD protocol. Such a naive protocol is implemented and there is an experimental study by means of a third-party urban fiber network QKD in Moscow [28].

4. QUANTUM INFORMATION AND BLOCKCHAIN

In this section we explain a possible solution by means of blind quantum computation, which is a quantum protocol to establish anonymous computation in a perfectly secured way. Blind quantum computation allows Alice who has a classical computer and a simple device to just generate a qubit to delegate her quantum computation to Bob who has a full-fledged quantum computer in such a way that he never knows any information of her input, output, and algorithm for quantum computation. This protocol is expected to protect users' privacy in a very near future when some quantum computers are available for commercial interests, and companies or governments with full quantum computers provide quantum cloud computing services for ordinary users. It is know that perfectly secured blind computation on a classical computer is very difficult and not suitable for practical applications, whereas quantum algorithm solves the problem in a very simple manner and several experimental studies are reported. We aim at considering its application to a blockchain architecture and making a desirable environment for users.

4.1 Quantum Computation

Before going to the detail, we shall start by explaining the bottom line of quantum computation. The general definition of quantum computation can be stated as follows:

(i) Prepare initial qubits (for example, $|0\rangle^{\otimes n}$).

(ii) Operate a unitary matrix U to the initial state $(U|0\rangle^{\otimes n})$.

(iii) Measure the state $(U|0\rangle^{\otimes n})$ and obtain results.

A general unitary matrix can be decomposed into several products of unitary matrices, called quantum gates, $U = U_m U_{m-1} \ldots U_1$ and a quantum circuit can be illustrated in Fig. 7.

In Fig. 7, time flows from the left to the right and U_1 operator acts on the first and the second states $|0\rangle \otimes |0\rangle$. If all operations complete, then measure new states $|\psi_1\rangle$, ..., $|\psi_4\rangle$, which ends a series of computation. This type of quantum computing is very similar to the classical computers today and called quantum gate computing. At present Google, IBM, Microsoft, and Intel lead the industry in this direction. From a different approach, D-wave has successfully developed a new type of quantum computer by means of the quantum annealing [11]. On one hand the universal gate quantum computers employ the superfluid, superconductor, or topological insulators; on the other hand the quantum annealing computer uses Ising model and it solves optimization problems such as the traveling salesman problem [29] by finding the minimum point of the energy (or Hamiltonian) of the form

$$H = \sum_{ij} J_{ij}\sigma_i\sigma_j + \sum_i h_i\sigma_i,$$

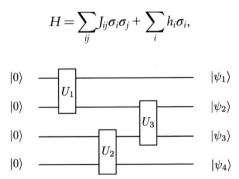

Fig. 7 Quantum gate circuit.

where J_{ij} is a component of a 2×2 matrix, σ_i takes values in $\{-1, 1\}$ which represent the spin of the atom at the ith sites, and h_i corresponds to a certain external magnetic field which is often used to describe the penalty term of the problem. Apparently and practically, quantum gate computers and quantum annealing computers are very different; however, it is known that they are theoretically equivalent in an ideal situation.

Let us introduce several fundamental matrices to explain more about quantum computation. We basically consider quantum gates. The following set of matrices are called Pauli matrices

$$X = \begin{pmatrix} 0 & 1 \\ 1 & 0 \end{pmatrix}, Y = \begin{pmatrix} 0 & -i \\ i & 0 \end{pmatrix}, Z = \begin{pmatrix} 1 & 0 \\ 0 & -1 \end{pmatrix}.$$

Readers may notice that they are unitary matrices ($XX^\dagger = 1$, for example). And we also invite the identity matrix

$$I = \begin{pmatrix} 1 & 0 \\ 0 & 1 \end{pmatrix}.$$

They satisfy the following equations:

$$X^2 = Y^2 = Z^2 = 1,$$
$$XY = -YX = -iZ.$$

Now define the expansion of $e^{i\theta A}$ with respect to a real parameter θ and a matrix A by

$$e^{i\theta A} = \cos\theta I + i\sin\theta A.$$

Then the θ rotation $R_x(\theta)$ around the x-axis is defined by

$$R_x(\theta) = e^{i\theta X}.$$

For example, $|0\rangle = \begin{pmatrix} 1 \\ 0 \end{pmatrix}$ is transferred to $R_x(\theta)|0\rangle = \begin{pmatrix} \cos\theta \\ i\sin\theta \end{pmatrix} =$ $\cos\theta|0\rangle + i\sin\theta|1\rangle$. Similarly, $R_y(\theta)$ and $R_z(\theta)$ are defined by

$$R_y(\theta) = e^{i\theta Y}, \quad R_z(\theta) = e^{i\theta Z}$$

and they act on $|0\rangle$ as

$$R_y(\theta)|0\rangle = \cos\theta|0\rangle - \sin\theta|1\rangle,$$
$$R_z(\theta)|0\rangle = e^{i\theta}|0\rangle.$$

Note that $|0\rangle$ is not rotated by R_z (only an overall factor $e^{i\theta}$, which does not contribute to computation, is generated). It is known that any n-qubits gate can be generated by the 3-tuple $\{R_x(\theta), R_z(\theta), CZ\}$, $CZ = |0\rangle\langle0| \otimes I + |1\rangle\langle1| \otimes Z$ (the Solovay–Kitaev theorem). One also should note that CZ commutes to R_z: $CZ \cdot R_z = R_z \cdot CZ$.

The things above are very fundamental for quantum computation and now we introduce more powerful quantum computation, called the measurement-based quantum computation (MBQC) [30] (Fig. 8):

(i) Prepare N qubits.

(ii) Measure the first qubit.

(iii) For the second measurement, compute the measurement direction classically.

(iv) Repeat the procedure above.

In the Z-direction, the arranged qubits form a graph G and the corresponding state $|G\rangle$ can be expressed by a unitary operator $|G\rangle = U|0\rangle^m$. The essential reason why the measurement of a graph state gives the corresponding quantum computation can be explained as follows. Suppose one has a qubit $|\psi\rangle = a|0\rangle + b|1\rangle$. Then operate the CZ gate with $|+\rangle$ and obtain

$$CZ(|\psi\rangle_1| +\rangle_2) = a|0\rangle_1| +\rangle_2 + b|1\rangle_1|-\rangle_2.$$

Project the first qubit to $|0\rangle + e^{i\theta}|1\rangle$ and as a result the second qubit changes into

$$a| +\rangle_2 + be^{-i\theta}|-\rangle_2.$$

Therefore, this operation is formulated as

$$|\psi\rangle \to He^{i\theta Z/2}|\psi\rangle,$$

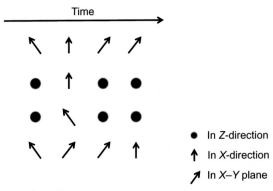

Fig. 8 Measurement-based quantum computation.

where H is the Hadamard operator

$$H = |+\rangle\langle 0| + |-\rangle\langle 1| = \frac{1}{\sqrt{2}}\begin{pmatrix} 1 & 1 \\ 1 & -1 \end{pmatrix}.$$

It is known that any 1-qubit state can be generated by the rotation operator R_Z and H, and it is possible to apply to more than 2-qubit states. The MBQC is a quantum computation done in a classical way; hence it allows one without a quantum computer to execute a command. This model is very important for blind computation and we will explain how to enjoy this quantum computation in the next subsection.

4.2 Blind Quantum Computation and Application to a Peer-to-Peer System

The blind computation is a way to execute a command sent by a customer (Alice) to a server (Bob) without knowing any information of Alice's input, output, and her algorithm. This protocol is crucial to protect Alice's privacy. Especially, when she makes a remittance, she will not want nobody (even Bob) to know to who and how much she tries to send money. We are mostly interested in making a privacy secured decentralized online banking system and to end we first explain what is the blind quantum computation. One of the strongest advantages in the quantum information theory is that it can realize the blind computation in a very simple way compared with the classical theory. Indeed, making a classical blind computation algorithm suitable for practical applications is still an open question. The blind quantum computation based on the BFK protocol [31] is performed in the following way:

(i) Alice sends N qubits $\{R_z(\theta_i)|+\rangle\}$, $(i=1,\ldots,N)$, where $\theta_i \in \{\frac{k\pi}{8} : k=0, \ldots,7\}$ are randomly chosen by Alice and keep secret to Bob.

(ii) Bob arranges the qubits on vertices V of a graph $G=(V,E)$ and gets

$$\left(\otimes_{i=1}^{N} R_z(\theta_i)\right)|+\rangle^{\otimes N}.$$

(iii) He operates the CZ gates to edges E and obtains

$$\left(\otimes_{e\in E} CZ_e\right)\left(\otimes_{i=1}^{N} R_z(\theta_i)\right)|+\rangle^{\otimes N}.$$

Since CZ_e commutes to R_z, it results in

$$\left(\otimes_{i=1}^{N} R_z(\theta_i)|G\rangle.$$

(iv) Alice and Bob do classical communication. Suppose the ith qubit is measured at states $R_{\phi_i}|\pm\rangle$. Then she sends $\delta_i = \theta_i + \phi_i + r_i\pi$, where $r_i \in \{0, 1\}$ are randomly chosen by Alice and keep secret to Bob.

(v) Bob measures the ith qubit using $R_{\delta_i}|\pm\rangle$ and tells her the result.

(vi) Alice receives the result and decides the next measurement angle by her classical computer.

(vii) Repeat the procedure above until she satisfies.

Bob has no chance to know any information of Alice if she hides θ_i and r_i to him. In this protocol, all she has to do is to generate quantum states by means of some instruments available today. Therefore the MBQC model opens up new possibilities of quantum computation without full-fledged quantum devices. In addition to that, simple quantum computers are unstable against noise in general and so-called the topological quantum computation realizes robust computation using geometrical properties of quantum codes [32]. And relevant blind computation protocols are proposed [33].

Now let us consider how to apply this protocol to a peer-to-peer system. The blind quantum computation may solve several problems discussed in previous sections in a very simple and powerful way. To address a realistic situation, we consider a society in which a full-fledged quantum computer is used for a central server and ordinary people commonly access it via classical channel as illustrated later. The central idea of us is to use the quantum server only for the blind computation so that participants with simple quantum devices on a peer-to-peer network can entrust their computation to the central server, and basically the main system itself should work without going through a particular third party. We explain a realistic application of this blind quantum computation to a peer-to-peer system. To utilize a quantum server, users send their very important information which they want to keep confident to it and write their open information into a blockchain. Sometimes such open information may include date or name. For example, if they want to shear several confidential documents among a community, they should use the blind quantum computation and share a private passcode corresponding to $\{\theta_i\}$ and $\{r_i\}$ in advance so that they can access to one of the documents, and write their name and date when they access to it into a blockchain, by which people in the community can check if a person who looked the document is exactly the one who is allowed to access it. Replacing a blockchain system for shearing medical record browsing history by this blind quantum system would ensure privacy of patients and improve the reliability of such a peer-to-peer system. Medical records usually belong to each hospital which takes full responsibility for their medical records and they are rarely passed to a different medical institute. Hence there exist many medical documents with almost the same contents created for the same patient in a different way and that reads to decrease the efficiency of health care services. Introducing a blockchain system for managing medical records in a unique way is proposed and indeed this strategy seems to be effective. However, not many use cases in clinical situations are reported since it requires special care and high techniques to create a completely secure blockchain in order to avoid privacy problems. Therefore, application of quantum computing to health care services in the proposed way may help to solve those issues.

The bind quantum computation will be also applicable to a cryptocurrency protocol. For example, CoinJoin anonymizes Bitcoin transactions by means of a central classical server. However, remember that such a central server should be trustworthy at all times; otherwise it has a chance to steal coins and disclose privacy of owners. To avoid trusting it, the CoinShuffle protocol was invented to implement CoinJoin in a decentralized manner. Still, complete anonymity is not achieved in a sense of classical computing. Therefore if one applies a quantum zero-knowledge proof to this apparently centralized system, one would realize a completely confidential transaction (Fig. 9).

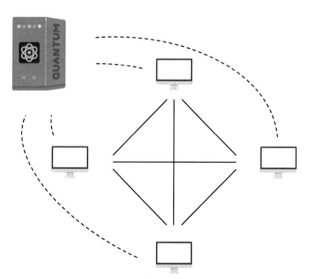

Fig. 9 A peer-to-peer classical network with a quantum server.

4.3 Quantum Money and a Peer-to-Peer Quantum Cash System

The concept of quantum money is very old and its history goes back to the invention by Wiesner in 1970. Now let us explain qBitcoin, a peer-to-peer quantum cash network which may be realized in the future quantum Internet [34]. Quantum states have very nice properties suitable for money. For example, making copies of quantum states by a unitary operator is impossible. This fact is known as the no-cloning theorem and can be explained as follows. Suppose we have different two pure quantum states $|\psi\rangle$ and $|\varphi\rangle$ and try to duplicate them by a unitary operation U. Making copies of them is formulated as follows:

$$U(|\psi\rangle \otimes |s\rangle) = |\psi\rangle \otimes |\psi\rangle$$
$$U(|\varphi\rangle \otimes |s\rangle) = |\varphi\rangle \otimes |\varphi\rangle,$$

where $|s\rangle$ is a unit state. Taking an inner product of them, we obtain $\langle\varphi|\psi\rangle = |\langle\varphi|\psi\rangle|^2$, which implies $\langle\varphi|\psi\rangle = 1$ or 0. Hence unless they are orthogonal, duplication of different quantum states is avoided. How can one send a coin made of quantum information to somebody else in a remote area? So-called the quantum teleportation gives a solution. We use the EPR pair defined previously. Let $|\psi\rangle$ be a coin which a remitter (Alice) wants to send to a receiver (Bob). Alice and Bob share an EPR pair and she performs a Bell measurement on one of the EPR pair $|\phi\rangle_{EPR1}$ and $|\psi\rangle$. Then she tells the outcome to Bob via a classical channel, by which the receiver can recover the information of $|\psi\rangle$ by performing a unitary operation on the other EPR pair $|\phi\rangle_{EPR2}$. Through this measurement, quantum states the remitter possessed are discarded and the coin $|\psi\rangle$ is sent to Bob. In this way, the coin data do not remain in Alice's hand and therefore this solves the double-spending problem. For successfully communicating the information of $|\psi\rangle$, Alice and Bob shear a private key using a QKD protocol. She decodes the outcome of a Bell measurement and tells it to Bob via open channel. Decoding it, he can get the coin (Fig. 10).

In order to trade quantum coins (qubit coins) in a peer-to-peer way, we introduce quantum digital signature, which is based on a one-way function

$$k \to |f_k\rangle \quad \text{easy}$$
$$|f_k\rangle \to k \quad \text{impossible},$$

where k is a classical bit string (private key) and $|f_k\rangle$ is a corresponding quantum state (public key). A well-known protocol is given by Gottesman and Chuang [35]. Their signature scheme is a one-time pad and requires some participants to verify signature; hence an owner (Alice) of a qubit coin needs

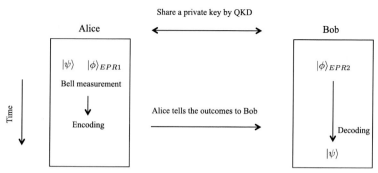

Fig. 10 Quantum teleportation to send a quantum coin.

to submit a signed transaction document to them. So the quantum digital signature is a peer-to-peer protocol initially. She should be careful not to distribute her public key to so many verifiers; otherwise her private key can be guessed easily.

The things very different from the classical cryptocurrency protocols are as follows. First, we address a peer-to-peer online cash system whose coins are given by a central mint. Each transaction can be completed without going through a particular third party. We chose this style for a reason of security. That is to say, the classical and traditional pure peer-to-peer protocols allow everyone to issue currencies as they wish, but that means there is a chance that insecure currencies are traded. Indeed, several cryptocurrencies suffer a great loss from a burglary and stolen coins never back to owners unless their serial numbers are traceable. In order to prevent such a situation, we assume a central mint. Second, when Alice sends her coin to Bob, she sends a quantum state with a signed transaction document and submits the document to signature verifiers without the quantum state so that they can confirm the transaction. If her signature is approved, then Bob is allowed to receive her coin. In this way this system solves the double-spending problem.

5. SUMMARY AND RESEARCH DIRECTIONS

We close this chapter with several insights into future research directions. In the first part of this chapter, we have discussed security and privacy issues of modern blockchain architectures based on several advanced researches into cryptosystems and classical algorithms. The author tried to explain advanced topics including recent affaires of cryptocurrencies in a simple manner, by which he expected that readers understood core objects successfully and clearly. Despite the relatively short history of decentralized systems, researches on this field have already achieved a good level and more fruitful results will be produced together with the quantum information science and technology, which were the main objectives in the latter part of this chapter. After all, it was a good surprise for the author to realize that describing the modern cryptosystems and the peer-to-peer systems from a view of quantum physics is a very good way to give an introduction to quantum information theory and technology. Though the materials prepared in this chapter were very fundamental, they covered the essential aspects of resent researches and will help readers to jump or begin anew in this field soon. As an orientation toward future studies, it would be interesting to consider applications of quantum information to a peer-to-peer system and there are still many things to be studied. For example, blind quantum computation, QKD, and quantum teleportation should be powerful tools and many applications to peer-to-peer network will appear not before long. Especially it will be meaningful to design a decentralized system on the quantum Internet. Developing a quantum currency like qBitcoin is one option but it is true that no one yet has gotten a clear view of a peer-to-peer system in the upcoming quantum era. It is a great pleasure for the author if readers of this chapter try to consider them and cultivate new fields by themselves.

KEY TERMINOLOGY AND DEFINITIONS

Blind quantum computation It is a secure protocol which enables a client (Alice) without a quantum computer to delegate her quantum computation to a server (Bob) with a full-fledged quantum technology in such a way that Bob cannot obtain any information about Alice's actual input, output, and algorithm. This protocol can become very secure and robust against realistic noise thanks to topological blind quantum computation. On the other hand, it is still an open problem to establish classical blind computation in a practical manner; thereby blind quantum computation is one of the advantages in generalizing a classical system to a quantum system.

Elliptic curve cryptography It is one of the most secure cryptography schemes in the sense of classical computation. This is based on the computational hardness of the mathematical problem associated with elliptic curves. (This is also an NP-hard problem.) In terms of blockchain, this is used for an open public key protocol and digital signature. However, it is pointed out that, if quantum computation technology is highly established in the next couple of decades, then elliptic curvature cryptography would be easily broken.

NP problems and NP-hard problems An NP problem is decision problem solvable in polynomial time by a theoretical nondeterministic Turing machine. Polynomial time refers to the increasing number of machine operations needed by an algorithm related to the size of the problem. NP-hard problems are informal defined by saying that they are at least as hard as the hardest problems in NP. For example, the traveling salesman problem belongs to NP-hard class and its problem size corresponds to the number of cities to visit.

Quantum cryptography and quantum key distribution Quantum cryptography is an alternative cryptography for classical cryptography as represented by the elliptic curve cryptography. This is based on quantum features of electrons or photons and its security is guaranteed by the laws of physics; thereby this protocol can be censure against any attack by means of any computer.

REFERENCES

[1] N. Koblitz, Elliptic curve cryptosystems, Math. Comput. 48 (1987) 203–209.
[2] V. Miller, Use of elliptic curves in cryptography, in: H.C. Williams (Ed.), Advances in Cryptology—CRYPTO '85 Proceedings. CRYPTO 1985. Lecture Notes in Computer Science, vol. 218, Springer, Berlin, Heidelberg, 1985.
[3] K. Ikeda, Quantum hall effect and Langlands program, arXiv (2017)1708.00419.
[4] E. Frenkel, Lectures on the Langlands program and conformal field theory, in: Proceedings, Les Houches School of Physics: Frontiers in Number Theory, Physics and Geometry II: On Conformal Field Theories, Discrete Groups and Renormalization, 2003.
[5] S. Nakamoto, Bitcoin: A Peer-to-Peer Electric Cash System.
[6] I. Eyal, E.G. Sirer, Majority is not enough: Bitcoin mining is vulnerable, ArXiv (2013) e-prints.
[7] R.L. Rivest, A. Shamir, Y. Tauman, How to leak a secret, in: C. Boyd (Ed.), Advances in Cryptology — ASIACRYPT 2001. ASIACRYPT 2001. Lecture Notes in Computer Science, vol. 2248, Springer, Berlin, Heidelberg, 2001.
[8] G. Maxwell, CoinJoin: Bitcoin Privacy for the Real World (Online). https://bitcointalk.org/index.php?topic=279249, 2013.
[9] T. Ruffing, P.M. Sanchez, A. Kate, in: CoinShuffle: practical decentralized coin mixing for Bitcoin, 19th European Symposium on Research in Computer Security, 2014.
[10] K. Grover, A fast quantum mechanical algorithm for database search, in: Proceeding STOC '96 Proceedings of the twenty-eighth annual ACM symposium on Theory of Computing, 1996, pp. 212–219.
[11] T. Kadowaki, H. Nishimori, Quantum annealing in the transverse ising model, Phys. Rev. E 58 (1998) 5355.
[12] F. Neukart, et al., Traffic flow optimization using a quantum annealer, Front. ICT 4 (2017) 29.
[13] M. Henderson, S. Adachi, Application of quantum annealing to training of deep neural networks, arXiv:1510.06356, 2015.
[14] M. Benedetti, et al., Estimation of effective temperatures in quantum annealers for sampling applications: a case study with possible applications in deep learning, Phys. Rev. A 94 (2016) 022308.
[15] C. Bennett, et al., Teleporting an unknown quantum state via dual classical and Einstein-Podolsky-Rosen channels, Phys. Rev. Lett. 70 (1993) 1895.
[16] D. Bouwmeester, et al., Experimental quantum teleportation, Nature 390 (1997) 575–579.
[17] A. Furusawa, et al., Unconditional quantum teleportation, Science 282 (5389) (1998) 706–709.
[18] S. Taleda, Deterministic quantum teleportation of photonic quantum bits by a hybrid technique, Nature 500 (2013) 315–318.
[19] H. Kimble, The quantum internet, Nature 453 (2008) 1023–1030.
[20] T.J. Berners-Lee, et al., World-Wide Web: Information Universe, Vol. 2 (No. 1), 1992, Electronic Networking (later renamed *Internet Research*).

[21] D. Aggarwal, et al., Quantum attacks on Bitcoin, and how to protect against them, arXiv (2017) 1710.10377.

[22] A. Einstein, B. Podolsky, N. Rosen, Can quantum-mechanical description of physical reality be considered complete? Phys. Rev. 47 (1935) 777.

[23] C. Bennett, G. Brassard, Quantum cryptography: public key distribution and coin tossing, in: International Conference on Computer System and Signal Processing, IEEE, 1984, pp. 175–179.

[24] A.R. Dixon, et al., Gigahertz decoy quantum key distribution with 1 Mbit/s secure key rate, Opt. Express 16 (2008) 18790–18797.

[25] M.A. Nielsen, I. Chang, Quantum Computation and Quantum Information, Cambridge University Press, 2010.

[26] P.W. Shor, Polynomial-time algorithms for prime factorization and discrete logarithms on a quantum computer, SIAM Rev. 41 (2) (1997) 303–332.

[27] L. Vandersypen, et al., Experimental realization of Shor's quantum factoring algorithm using nuclear magnetic resonance, Nature 414 (2001) 883–887.

[28] E. Kiktenko, et al., Quantum-secured blockchain, arXiv (2017)1705.09258.

[29] K. Ikeda, et al., Traveling Around Dream and Magical Lands With D-Wave Quantum Computer, 2018. to appear.

[30] R. Raussendorf, H.J. Briegel, A one-way quantum computer, Phys. Rev. Lett. 86 (2001) 5188.

[31] A. Broadbent, J. Fitzsimons, E. Kashefi, in: Universal blind quantum computation, FOCS '09 Proceedings of the 2009 50th Annual IEEE Symposium on Foundations of Computer Science, 2009, pp. 517–526.

[32] A. Kitaev, C. Laumann, Topological phases and quantum computation, arXiv (2009) 0904.2771.

[33] T. Morimae, K. Fujii, Blind topological measurement-based quantum computation, Nat. Comput. 3 (2012) 1036.

[34] K. Ikeda, qBitcoin: a peer-to-peer quantum cash system, in: 2018 SAI Computing Conference (SAI), 2018.

[35] D. Gottesman, I. Chuang, Quantum digital signatures, eprintarXiv (2001)quant-ph/0105032.

ABOUT THE AUTHOR

Kazuki Ikeda is a PhD student at Osaka University, Japan. He was selected to an Honorable Graduate Student of Graduate School of Science, Osaka University (2015–20). He was selected to a creator of MITO Advanced Program, which is a part of Exploratory IT Human Resources Project run by Information-Technology-Promotion-Agency, Japan (IPA) (2017).

His research interests include quantum physics, mathematical physics, information science (both classical and quantum), and artificial intelligence. His research papers are published in many conferences, workshops, and International Journals of repute including IEEE. He engaged in bootstrapping several international businesses.

CHAPTER EIGHT

Privacy Requirements in Cybersecurity Applications of Blockchain

Louise Axon, Michael Goldsmith, Sadie Creese
Department of Computer Science, University of Oxford, Oxford, United Kingdom

Contents

Abstract

Blockchain has promise as an approach to developing systems for a number of applications within cybersecurity. In Blockchain-based systems, data and authority can be distributed, and transparent and reliable transaction ledgers created. Some of the key advantages of Blockchain for cybersecurity applications are in conflict with privacy properties, yet many of the potential applications have complex requirements for privacy. Privacy-enabling approaches for Blockchain have been introduced, such as private Blockchains, and methods for enabling parties to act pseudonymously, but it is as yet unclear which approaches are suitable in which applications. We explore a set of proposed uses of Blockchain within cybersecurity and consider their

Advances in Computers, Volume 111
ISSN 0065-2458
https://doi.org/10.1016/bs.adcom.2018.03.004

requirements for privacy. We compare these requirements with the privacy provision of Blockchain and explore the trade-off between security and privacy, reflecting on the effect of using privacy-enabling approaches on the security advantages that Blockchain can offer.

1. INTRODUCTION

Blockchain technology has properties that could be advantageous in applications relating to cybersecurity. A number of uses of Blockchain within cybersecurity have been proposed in recent years, from ensuring the integrity of data storage in health care and finance to constructing emerging Internet-of-Things (IoT) networks and ad hoc networks. The properties provided by the Blockchain, which functions as a distributed ledger that can reliably preserve a record of transactions, are a natural solution to a range of cybersecurity problems.

In its original form, Blockchain is transparent insofar as posted and stored values are viewable by all who access the Blockchain. These values may by posted in plain sight, or encrypted, and are generally stored as hashes of the transactions in a particular block. Hashed values are calculated using one-way hash functions that do not reveal any information pertaining to the transactions (subject to brute-force attacks). By verifying that a hash of a given value is present at a particular time stamp in the Blockchain, a transaction's existence at that time can be proved. Merkle trees can be used to construct reliable hashes of a block's transactions, for example [1]. The distribution of this update and storage approach across Blockchain network members means that posted transactions are not private by default.

Several of the cybersecurity applications for which Blockchain has been proposed have specific and complex requirements for privacy. In securing health care data, for example, the types of data that should be viewable differ between entities on the system (such as patients, medical professionals, and regulating authorities) suggesting a requirement to control varying levels of read access to the data. In other applications, such as securing key infrastructures in mobile communications networks, the parties using the network require a level of privacy that prevents tracking of their actions across multiple locations, yet sufficient transparency must be retained to enable effective vehicular communications. These two examples illustrate the way in which the specific application of Blockchain technology to security leads to differences in the nature of the privacy required.

The balance between security and privacy in the use of Blockchain is complex. There is a juxtaposition, for example, between the transparency of Blockchain transactions, which are posted to the Blockchain to be seen by any party with the ability to view the Blockchain, and the possibility for any posts to the Blockchain to be encrypted: the contents of posts to the Blockchain are not prescribed and can in theory take any form. The fact that the Blockchain leaves a naturally reliable, auditable trail of transactions is desirable for ensuring the security of certain applications. In computer supply chain networks, for example, using Blockchain could enable the origin of different computer parts to be reliably tracked, in order to ensure the absence of counterfeit parts. It may be inappropriate, however, for all parties to be able to view all processes carried out by particular computer part vendors without restriction, and this is in conflict with the advantages of a transparent audit trail.

While the Blockchain has clear advantages in many use-cases, therefore, it clearly has the potential to be invasive to privacy if not applied in appropriate cases, or using suitable approaches [2]. Given the tensions between the security properties that can be given by Blockchain, and the privacy requirements of applications in which it has been proposed, it is imperative that an appropriate balance between security and privacy requirements is struck in applications of Blockchain. Exploring the trade-off between the security and privacy requirements of different Blockchain applications is important in producing appropriate designs for these applications. The privacy required in Blockchain applications must be specified, and mechanisms by which these levels of privacy can be met identified. In order to ensure that the security functionality of Blockchain applications is preserved when these approaches to privacy are used, security properties and privacy requirements should be considered in conjunction.

In order to explore the trade-off between these requirements, we examine particular examples of the broad range of proposed applications of Blockchain relating to cybersecurity. We do not claim to provide a review of all proposed Blockchain-based systems that have targeted applications in cybersecurity, but select those that we consider have interesting security and privacy requirements. Rather than providing a detailed analysis of the security or validity of particular systems, we consider the implications more generally for security and privacy of using Blockchain in the applications at which they are aimed. Our scope covers applications within data storage, management and execution of transactions (smart contracts), and networked communications. We select contrasting examples within these areas and consider their requirements. We explore the requirements for three case studies in depth: the electronic storage and management of medical records, voting using smart contracts, and networked communications for vehicular ad hoc networks/smart cities, extracting more detailed privacy requirements for these particular cases.

Some of the cybersecurity applications of Blockchain that we explore are strongly linked to privacy. An example is using Blockchain to store data in a decentralized way, in order to avoid its collection by large powers. The boundaries between our selection of security applications (which may be related to privacy in such a way) and our description of their privacy requirements therefore require clarification. In this chapter, we examine the extent to which it is possible to retain appropriate levels of privacy in the actual implementation of these security applications using Blockchain. The aforementioned case—the decentralization of data storage to avoid its collection by large powers—for example, could address wider privacy concerns relating to the storage and treatment of personal data. Dependent on its design, however, the decentralized data storage system itself could allow other parties viewing the Blockchain to read the data, which may also contradict privacy requirements. We therefore examine the types of privacy required in system design and explore methods by which they can be achieved.

In this chapter, we consider the properties of Blockchain, in order to scope a set of cybersecurity applications in which Blockchain has the potential to provide useful security properties, some systems for which have already been developed in industry and are in real-world use. For the scoped applications, we explore requirements for privacy-awareness and analyze the extent to which these can be met in Blockchain-based approaches, using existing mechanisms for privacy preservation. The requirements for privacy are illustrated in case studies of some proposed applications of Blockchain: storage of medical data, voting using smart contracts, and communications over vehicular networks. We consider the privacy provision of Blockchain against privacy requirements in these areas and explore the trade-off between security and privacy in building Blockchain applications.

The technical terms used in this chapter are explained wherever they appear. References are included at the end of the chapter.

2. CYBERSECURITY APPLICATIONS OF BLOCKCHAIN

2.1 Why Use Blockchain for Cybersecurity Applications?

Blockchain has a unique combination of properties that make it suitable for a number of applications. Proposed and existing applications include smart contracts, storage of sensitive data, and IoT device interactions. In Fig. 1, we illustrate the properties that are provided by a "default" Blockchain. We consider that this "default" Blockchain is one in its original, unaltered form, in which transactions (which may be encrypted) are posted by network members, verified by the network, and the resulting ledger of transactions is stored at network nodes, usually containing hashes calculated from the posted values. While approaches discussed later in this chapter can alter these properties, we view these as adjustments to the original form of a Blockchain and consider that the properties we describe in Fig. 1 are those provided by a Blockchain in its original form.

We clarify our description of the treatment of the data posted in this "default" Blockchain, presented in Fig. 1. We consider that data could be posted in plaintext. In reality, these values would often be hashed

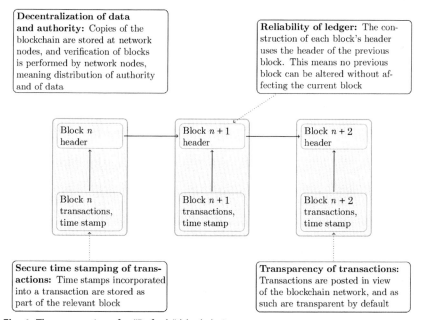

Fig. 1 The properties of a "Default" blockchain.

(or otherwise encrypted), with block headers (hash values calculated using a block's multiple transactions, and the header of the previous block, using Merkle trees) stored to be viewed by the network, retaining the reliability of the ledger and preventing tampering. Posting the data as plaintext gives the lowest level of privacy for that data. It is a possible design choice for a Blockchain-based system, however, and may be appropriate in some cases in which complete auditability of a system is more important than obscuring the information posted. Since the focus of this chapter is privacy, we take into consideration this "worst case" for informational privacy, in which this data is posted as plaintext.

As shown in Fig. 1, data and authority can be decentralized in Blockchain-based systems. This means that the data stored, and the power to make decisions on system processes, is distributed between multiple parties. Since the Blockchain is distributed between the network nodes held by the parties involved in the network, the storage of the Blockchain and its data is decentralized across these multiple nodes, such that the whole network stores the same copies of the Blockchain, rather than a subset of the network members. The verification process of Blockchain, in which all network members can compete in a system to verify and authorize new Blockchain transactions, means that the decision on the correctness of, and whether to accept, new transactions is not limited to any single powers, but is distributed between network members. There is therefore the opportunity for decentralization of both data and authority.

Because of the way Blockchain functions, it constitutes a reliable ledger of the transactions carried out over it. The posting of transactions, which are committed to time stamped blocks after verification, builds a record of all transactions carried out on the Blockchain. The calculation of each new block incorporates both new transactions and information relating to the previous block. A hash pointer is included in each block, which points to the previous block, and the hash of its header. This means that no block can be altered, without also altering the hash value, and rendering the hash pointer of the following block incorrect. This means that tampering with transactions recorded on the Blockchain in the past can be detected as changes to the value of the current state of the Blockchain. Past transactions are therefore tamperproof once logged.

Blockchain transactions are transparent by default, since posts to the Blockchain are made in the view of the rest of the Blockchain network, to be verified and committed. This means that transactions carried out can be seen in the clear, in the form of plaintext, or hashes, as discussed.

In reality, in most functioning Blockchain systems, this is not the case, since for privacy requirements to hide aspects of the data posted to the Blockchain have led to the use of privacy mechanisms such as the encryption of the data posted. In principle, however, a Blockchain-based system has the potential to enable the logging of transactions entirely transparent to all using it.

We now consider how these properties of Blockchain relate to cybersecurity applications. We explore the properties that are desirable in cybersecurity applications that can in theory be provided by Blockchain technology. The relations between Blockchain properties and cybersecurity applications are illustrated in Fig. 2.

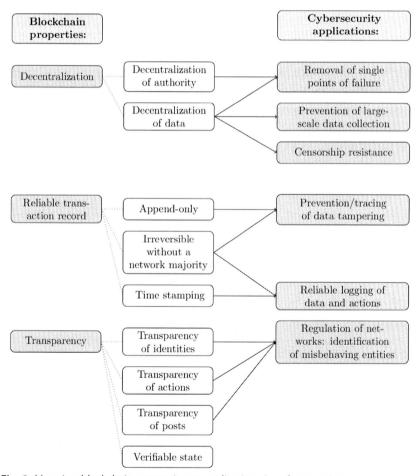

Fig. 2 Mapping blockchain properties to applications in cybersecurity.

The decentralization possible using Blockchain has different types. The decentralization of authority is shown in Fig. 2. This can be a desirable security property for some applications, enabling the removal of single points-of-failure controlling systems, who could be either owned by malicious entities or be insecure and become compromised by malicious entities, who then have power over systems. This reduces the need for users in a system to have trust in either a central party or governing authority, since the system can effectively be "governed" by the network of all its users.

Decentralization of data also has advantages for security. As with the decentralization of authority, decentralizing the locations at which data are stored removes the presence of single points-of-failure, at which data is stored and could be lost, stolen, or encrypted by ransomware, for example. If the data is stored at multiple nodes, this risk is reduced considerably, acting as a form of backup.

A key modern-day privacy concern that the decentralization of data is an approach to addressing is the prevention of mass data collection or processing by a small pool of organizations or entities. In general, there is public concern based on uncertainty surrounding the extent of the collection of personal data by large organizations, and the ways in which it is used or may be used in the future. The general lack of transparency in the storage and processing of such data can negatively affect public trust in the processes surrounding the collection of their data. There is a trend for people to want agency over their own data, such that it is not controlled by large powers, and its use is more transparent. For this problem, decentralizing some of the processes by which the personal data are collected and managed may present an attractive solution.

The Blockchain is a reliable and theoretically tamperproof ledger of transactions and assets that is distributed between and verified by a network of users. Blockchains leave a time stamped audit trail that cannot be tampered with secretly, and in this regard transaction records developed using Blockchain have strong security properties. Due to the time stamping of blocks, a proven record exists of the existence of a Blockchain ledger state at a given instance in time, meaning that the presence of particular transactions at certain times can be reliably audited.

The transparency of events on the Blockchain—the fact that copies of the Blockchain are accessible by all network members, and that actions carried out over the Blockchain are viewable by these parties by default—is also potentially advantageous in certain security applications. First, it could enable the regulation of networks: if all transactions carried out over a network are reliably and transparently logged, then the identification of misbehaving users is trivial. Furthermore, the transparency of the logging of processes carried out within systems, such as storage or processing of data, can give people confidence that they are aware of exactly what happens to their personal data, for example, and thus enable public trust in systems. Clearly, this transparency is in conflict with privacy of transactions over networks and in data storage, as discussed later, and a balance must be achieved.

2.2 Proposed and Existing Applications of Blockchain in Cybersecurity

In recent years there has been a surge in proposals for applications of Blockchain relating to cybersecurity. In the remainder of this section, we present a range of these applications for which Blockchain-based approaches have been proposed or developed, explore the approaches to enabling them using Blockchain, and consider the claimed advantages of doing so.

2.2.1 Securing Data Using Blockchain

Blockchain is a promising technology for secure data storage, because of the reliable ledger it creates. As we explained in Section 2.1, past transactions are tamperproof given an honest network majority, meaning that once verified and included in a Blockchain, the state of some data cannot be changed secretly. The ability to log time stamps reliably is another important property for data storage. Given that past transactions cannot be tampered with secretly, and that all transactions are logged, using Blockchain for the storage of data can instil confidence in those whose data is stored that they know what is happening to that data. Since the data is distributed across multiple nodes, it is secured against loss or encryption through ransomware, for example, since unless this occurred at every node, it could always be restored.

The decentralized nature of the storage systems Blockchain can create could be advantageous. The collection of data at single points—often large companies or governments—can mean that the collection, processing, and use of data is far from transparent. Using Blockchain, data can be stored in a decentralized way, in which actions on it are transparent across multiple nodes, and owners of the data (those to whom personal data belongs, for example) can have agency over their data, knowing where it is stored, and if it is modified in any way. In a post-Snowden culture in which there is general concern about the use of mass data collection and surveillance, this may be beneficial. It does, however, present clear concerns with respect to privacy. The storage of data at multiple nodes in a network, if not implemented properly, could mean the sharing of information with more parties, unintentionally, with whom this information should not be shared.

A range of systems for data storage using Blockchain have been proposed. Some of these systems have been generic in use; a data management system was proposed, for example, to allow users to control their own data in a system for storing, sharing, and querying data [3]. Such architectures could be applied to securely handle a wide array of different types of data.

In health care, the implementation of electronic medical records (EMR) is a key area of research. The use of EMR can augment the performance of health care services by increasing health care intelligence, improving the experience of patients in the health care system, and reduce running costs. The potential for Blockchain to transform EMR has been discussed [4]. Systems for EMR using Blockchain have been proposed in a number of works, with a view to exploiting the use of Blockchain as a reliable ledger to enable secure data storage and patient agency over their own data.

A particularly notable instance of the use of Blockchain in EMR is in Estonia, where there is a nation-wide adoption of Guardtime [5], a system based on Blockchain that is used to secure the health records of the entire country. Elsewhere, Ekblaw et al. proposed MedRec, a system that uses Blockchain technology to manage the security of electronic health records and medical research data [6,7]. Peterson et al. presented a Blockchain-based approach to sharing patient data in health care networks [8,9].

Yue et al. proposed an architecture for the Health care Data Gateway app, which uses Blockchain to enable patients to control and share their own health care data [10]. Secure multiparty computation, which we describe in Section 4.2.2, is proposed as an approach to enabling computation over patient data by third parties without compromising privacy, and an access control model is proposed to preserve privacy-awareness in access to data.

In financial sectors—banking and insurance, for example—there are potential benefits of using Blockchain-based technologies for data. Peters and Panayi outlined a number of applications for financial data storage using Blockchain [11]. The advantage of using Blockchain in this space, incentivizing banks, insurers, and financial institutions to use Blockchain technologies for data storage applications, is the potential to reduce overheads.

The regulation and auditing processes can be costly and time-consuming, and automating these processes further, using Blockchain technology to record financial transactions and execute processes, could improve the efficiency with which transactions could be processed.

Examples of such applications, in which the use of Blockchain technology in the recording and management of financial data has the potential to be useful, include transaction processing and banking ledgers [11].

Banking ledgers, and the recording of new financial transactions on them, could be automated in decentralized ledgers. The natural auditing property of Blockchain, and opportunity it offers for observation and detection of tampering, could serve as an auditing function, reducing the usual overheads this creates. Furthermore, the regulation of the behavior of banks could be bolstered by the observation of the public of the actions taken with their monetary accounts, in a decentralized and observable ledger.

2.2.2 Securing Transactions Using Blockchain

Smart contracts (sometimes called Blockchain contracts, self-executing contracts, or digital contracts) are programs that define the rules governing transactions. These programs allow parties to transact without the need for a trusted third party, since they are enforced by a network of peers [12]. Smart contract transactions are posted, and gossiped through peers before being eventually mined, and the transaction evoked.

They must include the functionality to execute like normal contracts, enabling mutual agreement of transactions and payments, the ability to enforce conditions, and the ability to withdraw under certain conditions. Smart contracts were first proposed by [13], and since then their use for devising and executing contracts—for example, in law, insurance, and crowdfunding—has been widely proposed. The use of Blockchain to build smart contracts is a natural fit, due to its decentralization. In Fig. 3, we illustrate the processes behind Blockchain-based smart contracts.

Blockchain-based smart contracts could enable secure contractual transactions or agreements in areas such as insurance, for example. Using Blockchain as the basis for smart contracts offers a number of advantages for securing the transactions made through contracts. The automation of tasks through software that would usually be carried out by middlemen such as lawyers and brokers means that transactions can be sped up, and users can have autonomy over the agreements they make, without relying on processes carried out by these third parties. This automation also means processes could be cheaper in theory, since middlemen are not required. Legal and transactions costs could be reduced, lowering the barrier for users to take part in a range of transactions [12]. If designed appropriately, the

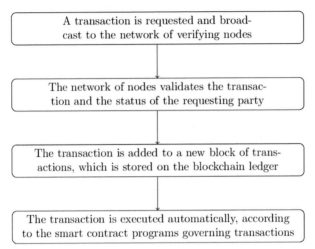

Fig. 3 Smart contract function.

automation of these processes could also reduce the opportunity for human error, such as errors filling out forms manually, improving the accuracy of transactions.

While advantageous in some ways, the automation of such processes produces new challenges and is potentially problematic in other ways. First, accidental bugs, or deliberate loopholes, in the code governing smart contract processes could lead to transactions that do not execute as understood, or intended, by the parties using them. Furthermore, automation could make the reversal of actions performed mistakenly by users difficult. In general, mistakes made by parties during traditional contractual processes could be rescinded in court. In smart contracts, however, transactions execute automatically, and deriving mechanisms for the reversal of payments made mistakenly and executed automatically poses a challenge.

Storing contracts on a distributed ledger could promote trust in the reliability of the transactions involved, given the reliability of the ledgers that Blockchain can provide for storing data: transactions cannot be tampered with or removed. The fact that the ledger is distributed among multiple nodes, and multiple parties own a copy of it, means that transactions cannot be lost or removed by single parties. In these ways, smart contracts are theoretically a reliable way of building tamperproof agreements between parties.

Smart contracts can be built on existing Blockchain platforms, such as Ethereum, a Blockchain platform for smart contract development [14].

A range of smart contract architectures and applications have been proposed, including legal contracts and businesses processes [15]. In the following, we explore a selection of these applications.

The use of smart contracts to improve data transparency in clinical trials has been proposed [16]. The aim was to improve scientific credibility of findings from clinical trials, which could be undermined by problems such as missing data and selective publication. Sreehari et al. proposed using Blockchain technology for drafting and probating wills [17]. The authors claimed that Blockchain technology has promise as an approach to designing wills that are tamperproof, secure, and transparent, and for increasing the speed of probation.

The Authoreon platform [18] aims to decentralize authorization, authentication, verification, and certification, combining Blockchain technology with cryptography and artificial intelligence. It is based the Ethereum platform [14] and uses use-case-specific smart contracts, combined with a database and file system. The platform uses an application-specific token, Autheon, for transactions. Authoreon will be offered as a platform, mobile app, Smart-Contract-as-a-Service, and plugin for third-party systems.

Blockchain protocols for voting have been proposed and are implemented in the Abu Dhabi stock exchange, for example, which is launching a Blockchain-based voting platform [19]. Blockchain-based *E*-voting has also been discussed recently in a report by the Scientific Foresight Unit of the European Parliamentary Research [20]. Proposed solutions to storing voting data on the Blockchain include FollowMyVote [21], The Blockchain Voting Machine [22], and TIVI [23]. In these solutions, the privacy of voters is achieved using a trusted authority, who obscures the relationship between a voter's voting key and their real-world identity, leaving them able to cast their votes to the Blockchain pseudonymously. These solutions therefore achieve voter privacy with the involvement of a trusted authority.

Smart contracts have been developed for boardroom voting. McCorry et al. proposed a Blockchain-based approach to boardroom voting, with the aim of ensuring privacy for voters by executing a voting protocol using the Blockchain consensus mechanism [24]. In that paper the Open Vote Network was proposed, a smart contract for Ethereum, designed to be suitable for boardroom elections. The implementation is self-tallying: no trusted authority is required to carry out the vote tallying or to protect the privacy of voters. All voting data is publicly available, yet the privacy of each voter, and information pertaining to that vote is controlled by that voter himself. The approach leverages the decisional Diffie–Hellman problem [25] and zero-knowledge proofs [26], to ensure that the individual vote cast by a particular entity can only be disclosed through the collusion of all the other voters. Unlike other solutions [22,23], therefore, this solution achieved voter privacy without the need to involve a trusted third party.

2.2.3 Securing Networks Using Blockchain

Blockchain presents a way of recording data. This data could be in the form of messages between parties, which are the basis of networks. Networked systems could in theory be built using Blockchain as a ledger of the communications between entities, and a system to enable entities to send messages to one another's accounts.

The decentralization offered by Blockchain has potential advantages for networks. It presents the opportunity to build networks in which no central authority is required, but which is controlled by the parties on the network, and their mining power. In cases in which it is more suitable to rely on an honest network than an honest central governing authority, this is appropriate. Relying on the honesty of a central governing authority may not always be desirable, since if the central authority is malicious (or is compromised by attackers and thus becomes malicious) control over the network could enable it to act adversarially: injecting false messages, rejecting or censoring certain messages, or eavesdropping on messages, for example. Furthermore, by reliably logging transactions in a tamperproof and time stamped way, using a Blockchain can prevent entities in a network from denying past transactions. Nonrepudiation of communications over the network can thus be achieved.

A number of applications of Blockchain for creating secure network designs have been proposed. This includes using Blockchain as the basis for messenger communications services. The Obsidian Blockchain platform and programming language [27] has been used to develop a secure messenger service, Obsidian Messenger [28]. The messenger service functions through nodes that are run by operators globally, using the Obsidian Blockchain. Node operators are rewarded transaction fees in exchange for transporting encrypted files and information [28].

The use of Blockchain for improving traceability in supply chain networks has also been proposed. Blockchain ledgers and time stamping can produce a reliable record of the time and location of production of goods, and identify when goods are fake, stolen, or have been diverted during transportation. An example is in improving the traceability of medicine, to prevent counterfeiting [29]. This also has implications for the traceability of supply chains in cybersecurity: improving the ability of end users and suppliers to trace the provenance of technologies such as computer parts, for example.

Communications networks generally use public key infrastructures (PKIs) to manage the entities' communications. PKIs are required for establishing identities and networking assets for networked communications, which enable parties to communicate using keys with which they can authorize transactions and prove their identity. The keys that entities use to establish communications channels are managed using PKIs. Blockchain has been proposed as a solution to constructing secure PKIs [30]. Blockchain can meet many PKI security requirements in theory and address some security problems of conventional approaches to building PKI. A ledger of PKI events can be published that is reliable as long as the majority of Blockchain contributors are honest [31].

PKIs are usually managed by Certificate Authorities (CAs) [32], who are responsible for issuing certifications of identity to network users. A security flaw in this approach is that CAs represent single points-of-failure. If a CA is compromised, so is the entire certificate chain for which that CA is responsible. This was evidenced, for example, by the high-profile 2011 hacking of CA DigiNotar [33]. Using a decentralized approach such as Blockchain means that these single points-of-failure are eliminated: the compromise of one single entity (as in the case of a CA) cannot lead to the subversion of the processes by which keys, and therefore trust in network members and their identities, are established.

Web-of-Trust (WoT) approaches to PKI are also proposed and used [34]. They are a decentralized method of building PKIs, in which entities are "trusted" if other entities already trusted on the network establish trust in them. However, WoT-based approaches have their own problems: they have a high barrier to entry in terms of trust [35], since they function by network entities becoming trusted by gaining the trust of other network users. In Blockchain-based PKI, this barrier to entry need not exist.

There have been various Blockchain-based PKIs proposed [36,37]. Wilson and Ateniese presented a Blockchain-based certificate format with which users can verify using Bitcoin identity-verification transactions [38], aiming to enhance the pretty-good-privacy model [39]. Some work has considered the development of privacy approaches for Blockchain-based PKI. Axon and Goldsmith proposed a Blockchain-based PKI with a focus on providing privacy-awareness for users [40]. The proposal drew on the Certcoin Blockchain-based PKI proposal by Fromknecht et al. [31], with restructuring to enable varying levels of anonymity for users.

A network that has emerged and grown rapidly in recent years is the IoT. We now consider the security properties that Blockchain technology can offer in this space. The IoT refers to the huge and increasing number of devices that have Internet connectivity. This includes wearable devices such as smart watches, smart home appliances such as fridges and televisions, and transportation modes in smart cities, for example. According to Gartner [41], there will be 20 million physical objects connected to the IoT by 2020.

Due to the huge, and rapidly expanding, scale of the IoT, with a multitude of connected devices varying greatly in nature and provenance, it poses a unique challenge with resulting difficulties of scale. To address the issue of large-scale communication and synchronization between the huge and rapidly growing number of IoT devices, the use of Blockchain-based approaches to reduce the many-to-one nature of traffic has been proposed [42], bringing potential advantages of efficiency and synchronization through the distribution of the communications ledger, and the decentralization of the verification process.

The decentralization of Blockchain could also address security issues such as the collection and storage of IoT data, and processing of IoT communications, at single points-of-failure, which could be malicious or compromised. A number of articles have proposed the use of Blockchain as a platform for the IoT, arguing for its potential advantages in this space [43–46]. We now explore some particular applications for the use of Blockchain within the IoT.

Dorri et al. proposed a secure Blockchain-based smart home framework [42]. In their approach, a single smart home has a "miner": a high-resource device that handles all communication between devices within the smart home, and with devices or entities external to the smart home. These communications are controlled and audited using a Blockchain. The authors claim that their approach is secure with respect to confidentiality, integrity, and availability. The use of symmetric encryption of communications preserves the confidentiality of messages sent between devices, while the hashing of messages maintains their integrity, since the hash stored at a particular time on the Blockchain can be verified to ensure the presence of that message at that time. The transactions acceptable by devices and the miner are limited, to achieve availability by preventing flooding or overloading of the system.

Huh et al. [47] proposed building IoT systems using Blockchain as a mechanism for controlling and configuring IoT devices. Their approach was based on using the Ethereum smart contract as a platform, to manage the configuration of IoT devices and build a key management system, with the storage of public keys for individual devices managed in the Ethereum Blockchain.

There are proposals for the use of Blockchains in vehicular networks. Vehicular networks manage communications between smart vehicles and are an integral part of the vision for smart cities. Rowan et al. leveraged both side channels and a Blockchain-based PKI, to develop an intervehicle session key establishment protocol [48]. Sharma et al. presented an architecture of a Blockchain-based vehicular ad hoc network for a smart city [49].

3. REQUIREMENTS FOR PRIVACY IN CYBERSECURITY APPLICATIONS

The level of privacy provided by systems is a key concern affecting the trust their users have in them, and the online safety of their users. There is a tension between the desire for privacy, and the need for traceability, in cases for law enforcement, for example. This results in a complex set of privacy requirements across applications within cybersecurity. In this section, we identify some of these types of privacy and position them against the applications to which they apply. We consider privacy requirements for data storage, smart contract transactions, and managing networks, illustrating the application of our privacy considerations through a case study in each of these areas: the storage and management of medical data, the use of smart contracts for voting, and the management of communications in vehicular ad hoc networks.

3.1 Types of Privacy

In many applications, a level of privacy is required, where we consider privacy to be the ability of the user to control their disclosure of information. The information that is referred to here can have different types, depending on the aspect of an entity's action that should be private. Based on the nature of the information referred to, we consider two types of privacy required in this chapter:

- *Privacy of information,* in which message content posted to the Blockchain is hidden. A party posting a message to the Blockchain may wish to hide the contents of that message from other network members.

- *Privacy of party*, in which identity of the party making a post to the Blockchain is hidden. If a party E posts to the Blockchain, E may not wish the rest of the network to know that he has posted at this time, or to be able to link this post to other posts by E.

Different levels exist within these two types of privacy. In many cases, complete privacy of an entity's actions toward all parties is not appropriate, and the ideal solution is complex, with varying levels of privacy required toward different parties. In the following sections, we outline the levels of privacy that could be required for both types of privacy: *privacy of information* and *privacy of party*.

3.1.1 Privacy of Information

By *privacy of information*, we refer to the extent to which the content of a post made by a party is known to the rest of the network. This information could be data for storage, a message to be communicated, or a request for a transaction, for example. We consider four different levels of *privacy of information*, which describe the extent to which information posted by a party is revealed to the rest of the network. These four levels of privacy are illustrated in Fig. 4, which illustrates the different levels of *privacy of information* a post P made by entity E_2 can have toward a set of other entities.

As shown in Fig. 4, a post P can be:

(a) *Visible to all entities (no privacy of information)*. If a post P is made by an entity E_2 to a Blockchain network consisting of the set of entities $\{E_1,...,E_8\}$, every entity in $\{E_1,...,E_8\}$ can view P.

(b) *Visible to a subset of entities (group-based privacy of information)*. If a post P is made by an entity E_2 to a Blockchain network consisting of the set of entities $\{E_1,...,E_8\}$, no information is disclosed about the contents of P to any entity other than the posting entity, E_2, and a subset of entities in $\{E_1,...,E_8\}$.

(c) *Visible to entities under certain conditions (conditional privacy of information)*. If a post P is made by an entity E_2 to a Blockchain network consisting of the set of entities $\{E_1,...,E_8\}$, no information is disclosed about the contents of P to any entity other than the posting entity, E_2, unless some condition is satisfied. In reality, such a condition may be a demand by a legal authority for disclosure of P, for example.

(d) *Visible to no other entities (total privacy of information)*. If a post P is made by an entity E_2 to a Blockchain network consisting of the set of entities $\{E_1,...,E_8\}$, no information is disclosed about the contents of P to any entity other than the posting entity, E_2.

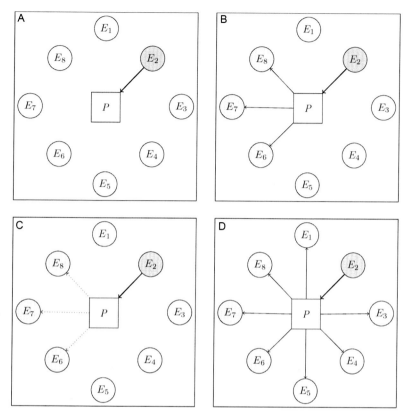

Fig. 4 Types of privacy of information. Post *P* made by entity E_2 has different levels of privacy toward other entities; entities to whom *P* is visible are shaded *light gray*. (A) Post visible to no other entities. (B) Post visible to a subset of entities. (C) Post visible to entities, conditionally. (D) Post visible to all entities.

3.1.2 Privacy of Party

By *privacy of party*, we refer to the extent to which the identity of a network member is known to the rest of the network. In the case of an anonymous entity *E*, where *E* has no presence in the network and makes transactions with no associated network identity of any type, the meaning *privacy of party* is simple: no other party knows any information whatsoever pertaining to the provenance of a post made by *E*. This is an extension of the *total privacy of party* level described later. In the case of pseudonymity, however, in which an entity *E* has a pseudonym *e* used for a network, there are different levels of *privacy of party* that can be achieved, which describe the extent to which the true identity of *E* is revealed by its actions.

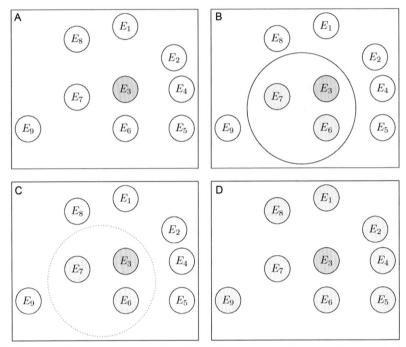

Fig. 5 Types of privacy of party. Entity E_3 has different levels of anonymity toward other entities; entities with knowledge of the identity of E_3 are shaded *light gray*. (A) Total privacy of party. (B) Neighbor-group privacy of party. (C) Conditional privacy of party. (D) No privacy of party.

We consider four different levels of *privacy of party*, which represent the extent to which information pertaining to the identity of an entity is revealed to the rest of the network. These four levels of privacy are illustrated in Fig. 5, which shows the different levels of *privacy of party* an entity E_3 can achieve toward a set of other entities. In Fig. 5, those entities who are able to obtain some information pertaining to the identity of E_3 are shaded in gray.

As shown in Fig. 5, an entity E_3 can have:

(a) *Total privacy of party.* If an action A is taken by an entity $E3$ under a pseudonym $e3$ on a Blockchain network consisting of the set of entities $\{E_1,...,E_9\}$, no entity in $\{E_1,...,E_9\}$ is aware that E_3 has taken action A, apart from the acting entity E_3.

(b) *Group-based privacy of party.* If an action A is made by an entity $E3$ under a pseudonym $e3$ on a Blockchain network consisting of the set of entities $\{E_1,...,E_9\}$, no entity is aware that E_3 has taken action A, apart from the acting entity, E_3, and a selected group of entities in $\{E_1,...,E_9\}$.

(c) *Conditional privacy of party.* If an action A is made by an entity $E3$ under a pseudonym $e3$ on a Blockchain network consisting of the set of entities $\{E_1,\ldots,E_9\}$, no entity in $\{E_1,\ldots,E_9\}$ is aware that E_3 has taken action A, apart from the acting entity E_3, unless some condition is satisfied.

(d) *No privacy of party.* If an action A is taken by an entity $E3$ under a pseudonym $e3$ on a Blockchain network consisting of the set of entities $\{E_1,\ldots,E_9\}$, every entity in $\{E_1,\ldots,E_9\}$ is aware that E_3 has taken action A.

The action A described earlier may consist of a post made to the network, or a verification carried out, for example.

3.2 Privacy Requirements for Securing Data

It is essential, if data is stored using a Blockchain-based system, that mechanisms are in place to prevent the disclosure of data to unintended parties. It may be not only the data itself that is sensitive, but details of actions linked to it, such as access and updates to it.

In an EMR system where each new piece of health information relating to a patient is posted to the Blockchain encrypted and can be linked to the patient in question, for example, there would be a privacy violation since other parties could view the frequency of updates made to that patient's information, which could be an indication of their health or number of visits to a medical practice. In this particular case of data storage, therefore, simply encrypting the data itself is not enough. In such applications, it is necessary that access to particular data is restricted to particular entities, and that these entities have varying types of access to the data—read or update access, for example.

In other types of data storage, there are parties from whom it is unnecessary to keep posts private—the medical records of a patient in the above example need not be kept private from that patient's medical team. There are cases in which it is inappropriate for data storage to have any level of privacy: the storage of electoral roll information, for example, traditionally uses an opt in privacy model, in which parties can choose to hide their electoral roll data, which is otherwise made public.

We now explore in more detail the privacy requirements for the particular case of storing medical data.

3.2.1 Case Study: Privacy Requirements for Medical Records

In extracting the privacy requirements for securing medical data using Blockchain, it is important to consider the benefits of using Blockchain in this application in the first place. Key advantages that Blockchain could theoretically offer for the storage and processing of medical data records are as follows:

- Strengthening patient trust, by enabling patients to access a transparent and reliable view of the ledger recording the uses of their own medical data.
- Creation of reliable and verifiable audit trails for the medical data stored for patients, or used in medical research.
- Improving the efficiency of processes involving medical data, such as dynamic patient consent processes and sharing of data in certain cases (in cases of medical emergency, for example), through automated execution of such processes under certain conditions.
- Opportunity for decentralized storage of data in different states across distributed nodes, such that no node is a single point-of-failure for all data, and data cannot simply be lost or removed by the compromise of a single node. This may also present privacy challenges: the creation of more, smaller points-of-failure, the compromise of which would lead to the disclosure of some data records.

For particularly sensitive types of data such as health care data, privacy violations are a great concern. There is little sense in developing public Blockchains that function through public verification for the storage of medical records. Decentralizing the storage of medical data records beyond the scope of the health care provider and its patients offers little advantage for patient trust, and would not be appropriate given the sensitivity of the medical information recorded. Blockchains private to particular health care providers are better suited to this application. Creating Blockchains private to particular networks of health care providers could improve the reliability and trustworthiness of the actions carried out by any single health care provider, by enabling other health care providers to audit these processes.

We therefore consider the privacy requirements for medical data records, within the scope of Blockchains private to particular health care providers, or networks of health care providers. In schemes such as EMR, it is ideal that patients have agency over their data, with the ability to view actions taken on it, control access to it, and consent to processes using it (in fitting with dynamic consent processes). It is important to implement mechanisms that ensure only intended parties are able to access any particular details of the health of the patient are essential. There is clearly a requirement for a level of *privacy of information*. We now consider the levels of *privacy of information* that may be required for EMR, using the levels we defined in Section 3.1.1.

Generally, in medical record handling, it is inappropriate for posts by a party to be visible to all entities, given the sensitivity of the information concerned. Cases in which posts by a party should be visible to no other entities under any conditions are also unlikely to arise. Rather, the most appropriate level of *privacy of information* lies somewhere in the middle, in which parts of a patient's EMR are visible to a subset of entities. It would be suitable, for example, for a patient and the patient's medical team to have access to that patient's medical records.

It may also be required that aspects of a patient's medical record be visible to entities under certain conditions. There are times at which it is ideal to disclose parts of a patient's medical record to entities who would not have access to this information by default. An example of such a case would be disclosure of the medical records of a person involved in an accident to the paramedics responding to their emergency call.

Given the sensitivity of health care data, in many cases it is desirable that not only the contents of posts on particular patients' health care are private (by encrypting the post containing the health information, for example), but that details of the time and number of new posts relating to a patient's health are not completely transparent. This means that there is a requirement for some level of *privacy of party*.

The main advantage of using Blockchain for EMR in terms of audit is that it enables other entities to check the reliability of the data stored on the ledger, and to verify its correctness before it is committed to the ledger. Using such a ledger, it can be verified at the time of posting, and retrospectively, that medical data were posted by the correct entities—the patient's doctor, for example—were not tampered with, and were not used in processes that contradict the consent of the patients. It is therefore desirable that the parties posting to the Blockchain do not have total privacy of party: that there is some opportunity to use the audit trail advantage given naturally by using Blockchain in this application.

Group-based privacy of party in this case would refer to those entities "close to" the patient whose medical data is posted having the knowledge that data on that patient is being posted. This closeness in this context might mean the medical team associated with a patient. Generally, a doctor would have access to the medical records of their patients, would make updates to them, and be able to access updates to them. It is therefore appropriate that the doctors of a patient are aware when their patient's medical record is updated and are able to link the updates to that patient. Enabling the patient to also view the identity of all entities who take any actions involving their medical records will be an important part of improving patient trust in medical data processes, and in consent processes for medical procedures and use of their medical data.

If *group-based privacy of party* were used, it would be important to consider the size of the group used. Similar to the case of *no privacy of party*, *group-based privacy of party* that used a very small group, consisting perhaps only of a patient and his doctor, would lose the auditability advantage of using the Blockchain: the fact that other network members can verify the correctness of those making and updating data on patients. It may therefore be appropriate to include impartial network members, or authorities within the health care provider, in a patient's group, in order to strengthen this auditing.

Another possibility would be to implement some *conditional privacy of party* mechanism, in which others—perhaps the regulators or authorities for the health care providers, are able to access the audit trail of entities making changes to the record of a patient, under certain conditions, such as if that patient believes there has been some misbehavior. There are legal cases that would also require some disclosure mechanism. In case of a court order to disclose a patient's medical records for a legal inquiry or similar, it is important that there is some mechanism to obtain an audit trail for the records of a patient.

It would be inappropriate to have *no privacy of party*, given the sensitivity of medical information. For example, it would not be suitable for all network members to be able to access information on the identities of those making posts to the ledger. This would inform parties of the fact that a patient's medical records were being updated more regularly than usual, which might be an indication of a change in the state of their health, and therefore a leakage of information pertaining to that patient's medical state.

3.3 Privacy Requirements for Securing Transactions

The use of smart contracts using Blockchain to execute and maintain a record of transactions has been proposed for a range of applications requiring contracts. Dependent on what these contracts are applied to, the privacy requirements differ. As we show in the Section 3.3.1, if used for voting, smart contracts do not necessarily need to preserve *privacy of information*, and indeed it would not be suitable for them to do so, as long as the pseudonymity of voters was maintained.

The privacy of smart contracts has been considered, and methods of preserving appropriate privacy in smart contracts proposed. Kosba et al. claimed that existing Blockchain-based smart contract designs lacked transactional privacy, since details of the transactions—agreements or amounts transacted, for example—were visible on the Blockchain. They proposed Hawk, a smart contract system in which transactions are stored so that they are not visible on the Blockchain [50].

3.3.1 Case Study: Privacy Requirements for Smart Contract Voting

Smart contracts could be used to facilitate voting, collecting the votes cast by parties, and carrying out automated calculation of the voting tallies to produce a result. In the traditional voting model, parties cast their vote to be viewed in plain sight by those counting the votes on a piece of paper, and the total number of votes each way is known publicly. The only thing obscured is the identity of the person who made each individual vote, yet it is important that no entity can cast a vote twice. E-voting systems can emulate this model by ensuring pseudonymity of voters, but leaving information pertaining to the votes cast in plain sight. We consider the meaning of these privacy requirements in a Blockchain-based voting system.

It is important to preserve the benefits of observation for Blockchain-based voting. By making this process transparent, such that the public can audit it, the democracy of voting processes can potentially be strengthened. If there were to be corruption of the voting processes by a government, for example, in which the voting counts were tampered with, a reliable ledger of votes made viewable in plain sight could be observable and audited by the public, enabling the detection of tampering, and bolstering public trust in the voting process.

It is therefore suitable in this application to have *no privacy of information*, if that information consists only of a vote: the selection of candidate made in an election, for example. While it is appropriate to have *no privacy of* information, there is a clear requirement for a level of *privacy of party*. In voting applications, the votes cast by an individual entity must be totally anonymous: there must be no means of tracing the vote cast by an entity back to that entity. The vote cast by an entity must disclose no information whatsoever about their identity to any other party, and there is therefore a requirement for *total privacy of party*. In the Ethereum smart contract platform, for example, the concept of an identity is a transaction address.

It is necessary, however, that no entity is able to cast a vote twice. Using a single pseudonym (key) for each voting entity, to prevent the recasting of a vote using another key, is a solution to this. This gives pseudonymity: a party's vote is cast, revealing their pseudonym, but not their true identity.

Another possibility is to use linkable ring signatures, as described in Section 4.2.3, in which no concept of identity (e.g., a pseudonym) is revealed at the time when a party's vote is cast, but the detection of a second post using the same key can be detected.

In ensuring that *total privacy of party* is maintained, it is important to ensure that information, which need have *no privacy* and could be stored in plain sight, does not reveal any information about the party who cast it. The time stamping of transactions, for example, could reveal some information about the provenance of a vote, if another party knows the time at which a person voted, or makes deductions about the times at which certain types of people (e.g., those who work at certain times of day) are likely to have voted. It is key that no party can link voters to their votes based on time. Mechanisms to ensure this, such as obscuring the time stamps of votes, inserting some random time delay into the times at which votes are posted to the Blockchain, or causing updates to occur in groups, at which multiple votes are posted to the Blockchain at time intervals, are required.

In summary, for voting using smart contracts, there is a requirement for *no privacy of information* and is not desirable, as long as parties are able to cast their votes anonymously.

3.4 Privacy Requirements for Securing Networks

For networks, it is important that privacy of communications is not compromised. In general, the actions of entities communicating over the network should not compromise their privacy by revealing information about their communications activity. The aspects for which privacy needs to be considered include:

➢ The content of messages
➢ The times of, and entities involved in, communications, and
➢ The location of individuals (i.e., communications networks should not enable tracing of the locations of entities using them).

There are requirements that contradict the above, however. For the regulation of certain types of networks, it is necessary that there is some capacity to trace misbehaving entities and remove entities from networks. These are examples of actions that may need to be taken by law enforcement bodies in vehicular networks. In private communications networks in companies who place restrictions on the content of communications, such capabilities might also be required.

There are types of network in which the privacy required varies toward different entities. In a private network for a company, for example, there may be restrictions, based on security or general policy, on the information that can be communicated over the company network. It is therefore required that the entities regulating the communications over the company network have read access on communications over the network and are able to trace the identity of the entities communicating for cases of misbehavior. For other users of the network, however, read access would be inappropriate.

In ubiquitous computing and the IoT, the requirement for *privacy of party* is particularly high. A user may interact with a computing system through multiple devices, such as laptops, smartphone, and wearable devices. If linked, the use of these devices could enable tracing of a user's actions and locations. In many cases, privacy is required such that the user's actions cannot be linked across multiple devices. This is also the case in ad hoc networks, in which the locations in which a single device is used changes. Use of the same device in multiple locations should not enable tracking of a user's whereabouts.

3.4.1 Case Study: Privacy Requirements for Vehicular Networks

We illustrate the above requirement for prevention of linking of interactions across multiple locations, using vehicular ad hoc networks as an example. PKI is required for secure intervehicular communications, but its use must not enable remote tracking of a vehicle's actions. The identity corresponding to a public key must not be publicly disclosed, or keys linked at update. Vehicular networks have particular requirements. It is important that tracking of vehicles should not be possible remotely using their communications, yet it is futile to prevent others from tracking vehicles by communications in cases where they could be doing so physically anyway—for example, if they remain within line of sight for an extended time period on a motorway (these requirements are illustrated in Figs. 6 and 7). This is a case where group-based anonymity could be an appropriate privacy level.

We begin by considering the security and privacy requirements for vehicular ad hoc networks. In emerging vehicular ad hoc networks for driverless and intelligent cars, there is a requirement for cars to communicate between one another, and with roadside units which provide communications support, in order to receive instructions and updates as to the status of their surroundings (e.g., road works), and to communicate their actions

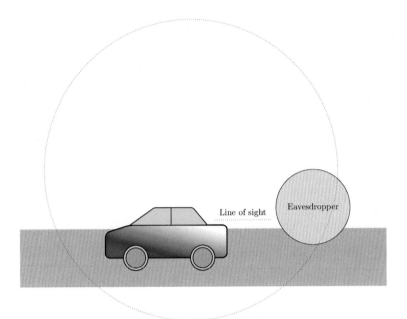

Fig. 6 Preventing tracking of vehicles by their VANET communications is futile for entities who are within line of sight.

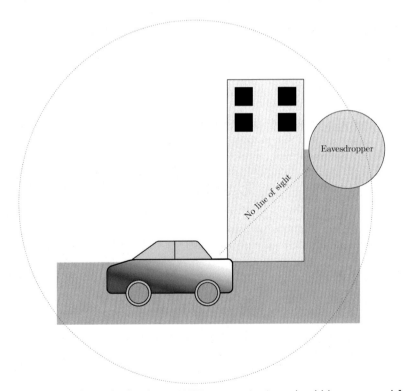

Fig. 7 Tracking of vehicles by their VANET communications should be prevented for entities not within line of sight.

(e.g., that they are planning to brake) to other vehicles. This drives a requirement for communication between vehicles.

Intelligent transport systems (ITSs) have a number of security requirements, falling into five categories: confidentiality, integrity, availability, accountability, and authenticity [51]. Broadly, security properties for ITS safety messaging arise from two main high-level requirements—safety and privacy—which are in many ways conflicting in this context: for safety applications, overtaking warnings, for example, we require that messages are sent frequently, that their contents are accurate, unaltered, and trustworthy, and that they are received by all those who require the information they contain. We therefore require that a message's sender is, and its contents are, trusted by its recipients. Privacy requirements, however, drive a requirement for recognition of message senders' true identities to be unobtainable from the safety messages, and for no opportunities for tracking vehicles to be given by safety messages. For safety messages, we identify five main areas in which security requirements arise.

In order for a message's sender and contents to be trusted, there is a need for authentication of messages. This means that a party should be recognized as a vehicle that is permitted to communicate over the network and must therefore have some presence in the network: *total privacy of party* is not suitable. Furthermore, the contents of safety messages must be viewable, and their integrity verifiable, by all vehicles affected by them: this may be those who are close by, and need to receive the information that a vehicle in their vicinity is braking, for example. It is therefore ideal for safety messages to be viewable as plaintext to at least a subset of parties on the network: *total privacy of information* is not appropriate.

For privacy, we require that opportunities for tracking are not given to attackers through the safety messages. According to ETSI "it should not be possible for an unauthorized party to deduce the location or identity of an ITS user by analyzing communications traffic flows to and from the ITS user's vehicle"; and "it should not be possible for an unauthorized party to deduce the route taken by an ITS end user by analyzing communications traffic flows to and from the ITS end user's vehicle" [51].

Fundamentally, we cannot achieve, using communication protocols, the prevention of tracking through physical tailing; it is therefore futile to include this prevention in the security requirements for the safety messages: it is always possible for a vehicle to be linked by a nearby party, within line of sight. It is therefore not required that two vehicles who remain nearby one another be unlinkable for that period of time since, trivially, the same tracking capability could be given in such a case by physical following of one vehicle by another. We illustrate this in Figs. 6 and 7.

What we require is that parties that are not nearby a given vehicle V are unable to track V through its safety messages (long-distance unlinkability), while those in some "nearby" vicinity are (short-distance linkability). Furthermore, we require that parties leaving the defined nearby vicinity of V for a given time cannot continue to link its safety messages on return to its vicinity to those it received before leaving its vicinity. Given these requirements, it appears that linkability properties based on distance may be more fitting than time-based ones.

Using Blockchain for vehicular ad hoc network communications could offer advantages. The opportunity to decentralize access to communications information over these networks is potentially advantageous. This could prevent any single party from viewing all information and thus tracing the actions of parties, and could mean that no single party is responsible for the storage and handling of this information, representing a single point-of-failure.

The case of ad hoc networks is somewhat different to the case of health care data considered in Section 3.2.1: there is a greater challenge in implementing ad hoc networks over fully private Blockchains. While it can be assumed that health care data could largely be kept within certain health care providers and shared to other health care providers only in rare cases, the case of vehicular ad hoc networks is much broader, and the boundaries between groups of entities using these networks are less clear.

It could be assumed, for example, that for the majority of the time a vehicle would operate within the physical boundaries around the places most often frequented by its owner: the home and workplace of the owner. Under this assumption, vehicular ad hoc networks over Blockchains private to certain districts, or cities, could be implemented. In reality, however, it is likely that a reasonable number of vehicle owners would travel to areas outside of their "usual" traveling zones each day—for work meetings with new clients, or leisure trips, for example. The implementation of such private Blockchains is complicated by these cases. It is likely more appropriate, therefore to build vehicular ad hoc networks using public Blockchains. This creates privacy challenges, however, in handling the data communicated by vehicles, and increases the pool of parties with the potential to track vehicles based on the timings and locations of their posts.

We consider the requirements for *privacy of information*. For safety messages, it is desirable that there is little *privacy of information*, since on the whole, the more information on surroundings and actions of vehicles that is available to other roadside infrastructure, the more detailed and correct safety decisions can be made based on this information. For a certain set of information, then, there is a requirement for *no privacy of information*.

Such information would include the fact that a vehicle has made the decision to brake. There is no need to hide this information from any party, if the identity of the vehicle has the appropriate *privacy of party* guarantees. It should not matter if an entity in a completely different city, City B, learns that a certain vehicle in a particular location in City A has decided to brake, but does not learn any information pertaining to the identity of that vehicle. For certain safety messages, we conclude that there is no requirement for *privacy of information*.

There is, however, a clear requirement that the information discussed earlier does not reveal the identity of the vehicle, and we now discuss the *privacy of party* requirements relating to this. It is important that information communicated over vehicular ad hoc networks should not enable the remote tracking of the movements and actions of a vehicle by adversarial parties. *No privacy of party* is therefore inappropriate in this case. Equally, there is some requirement for verification of the identity of those making posts on the behalf of vehicles: adversaries should not be able to make posts, posing as another vehicle, and misleading other parties as to the whereabouts of that vehicle, for example. The communications made by vehicles should be signed in such a way that the correctness of the posting vehicle can be authenticated. For this reason, *total privacy of party* would also be inappropriate.

It is futile to prevent tracking of a vehicle through its communications, by other parties who are within line of sight of that vehicle (vehicles that remain in close proximity for a time while driving on a motorway, for example). Such parties would be capable of physical tracking of that vehicle anyway, as we illustrate in Figs. 6 and 7. *Group-based privacy of party* could therefore be appropriate in vehicular ad hoc networks. These groups could be formed based on line of sight, or proximity to vehicles, using some distance-bounding protocol, for example. In such a way, a level of *group-based privacy of party* could be achieved, in which those parties (vehicles and roadside units) who fall into the group within a certain distance of a vehicle can view and attest the correctness of its actions, while others cannot, and hence cannot track the movements of that vehicle remotely.

In cases of misbehavior in vehicular networks, there is a requirement for tracing of entities. This could mean a party on the vehicular ad hoc network sending out misinformation, for example, or flooding the network with messages to deny service. In these cases, there is a need for some *conditional privacy of party*. Authorities for the vehicular network, which may be legal or policing authorities and are often termed Tracing Managers, may require access to vehicular communications records, in order to trace and verify the misbehavior of a party, and take appropriate actions. In these cases, it would be required that the actions performed by a particular vehicle on the network could be traced under certain conditions, such as at the request of the police for the records of that vehicle. The nonrepudiation property inherent in the ledger produced by Blockchain is advantageous in this case.

4. DEVELOPING BLOCKCHAIN APPLICATIONS WITH APPROPRIATE PRIVACY PROVISIONS

The proposed cybersecurity applications of Blockchain outlined in Section 2 have a range of requirements for privacy. The aim of this section is to explore the privacy provision of Blockchain, and to compare this with the privacy requirements we established in Section 3, for the cybersecurity applications considered. The idea of privacy on the Blockchain is complex and somewhat contradictory. In building any new mechanisms into Blockchain systems, it is ideal to preserve the security benefits of the Blockchain: the processing of transactions by a decentralized network, and the construction of a reliable transaction ledger that can be verified by all parties. Yet, while these security benefits take root in the openness of the Blockchain—its distribution and transparency—for fully privacy-preserving applications, the underlying information and parties involved in processes should be entirely obfuscated.

We begin by exploring the natural privacy provision of Blockchain, considering the "default" Blockchain introduced in Fig. 1. We then review proposed and existing approaches to altering the types of privacy that can be achieved using Blockchain-based systems.

4.1 Natural Privacy Properties of Blockchain

In its "default" form, Blockchain is transparent and functions through the confirmation of transactions seen by its network members. This means that transactions posted to the Blockchain are not private by default. In Fig. 8, we illustrate the level of privacy present in this "default" Blockchain.

Many of the theoretical advantages of using Blockchain in cybersecurity applications take root in the transparency of the system, and the distribution of the ledger to enable consensus-based verification of the correctness of

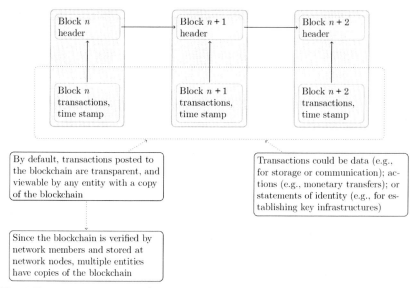

Fig. 8 The privacy provision of blockchain by default.

transactions: properties that we discussed in Section 2.1. Another key advantage of Blockchain is that it enables fast transactions and execution of processes over data that is shared between parties [52]. These advantages are in some ways in conflict with a number of general privacy requirements. In the "worst case" for privacy in the "default" Blockchain we described in Section 2, in which all parties can view posts to the Blockchain, and these posts are made in plaintext, there is no privacy of information or party.

The transparency of the system in its simplest state, in which posts to the Blockchain are visible to all parties viewing the Blockchain, means that all information posted to the Blockchain is not private. Any party who can view the block is able to read all the values stored on it, whether they are hashed, encrypted, or stored as plaintext. Clearly, for many applications, including the storage of sensitive data, this lack of informational privacy is not suitable. Furthermore, using Blockchain leaves behind a chain of information that can be audited, to check for correctness. This makes it a promising platform for security processes in which the presence of a reliable ledger of the past transactions is important, yet the fact that a trail is left by any transactions posted to the Blockchain, which can be audited at any time, contradicts transactional privacy.

The distributed nature of the Blockchain—the fact that it is stored at, and verified by, multiple Blockchain network nodes—means that the parties who own copies of the Blockchain and are thus able to view it can be very large. In public Blockchains, there are no restrictions on these parties: in theory, any party could view the Blockchain. This, in combination with the transparency of the posts we considered earlier, means that posting information to a Blockchain in its default form is effectively broadcasting that information in plaintext for anybody in the world to read. While this could improve the efficiency of many processes, enabling automated execution of transactions, and verification of the actions of parties by other parties computing over shared platforms, this sharing of information has huge implications for privacy.

Thus, while the advantages of Blockchain can be compelling for many applications, the resulting potential privacy implications cast doubt on whether it is suitable to use Blockchain in many cases. For some applications, companies and individuals are reluctant to post their information to a public database, to be read without restriction by any desiring party:

> *"As seductive as a Blockchain's other advantages are, neither companies or individuals are particularly keen on publishing all of their information onto a public database that can be arbitrarily read without any restrictions by one's own government, foreign governments, family members, coworkers, and business competitors."*
>
> **Buterin [52]**

In order to be suitable for adoption in many applications, it is imperative that guarantees of appropriate levels of privacy in Blockchain-based systems are achievable. This is particularly true in cybersecurity applications of Blockchain, since in many applications within cybersecurity, the information published is sensitive. In Section 4.2, we explore the range of approaches that have been proposed for building levels of privacy into Blockchain-based systems.

4.2 Approaches to Developing Privacy in Blockchain-Based Systems

We explore proposed and existing methods of constructing levels of privacy into Blockchain-based systems. We begin by exploring the idea of permissioned Blockchains, in which only certain permissioned parties can perform verification for a Blockchain, and Blockchains private to certain groups of parties. We then explore methods of hiding information from even those viewers who are able to access the Blockchain, and of obscuring the identities of posters to the Blockchain.

4.2.1 Controlling Access to the Blockchain: Private and Permissioned Blockchains

There are two models for controlling access to the Blockchain: private or public Blockchains, and permissioned or permissionless Blockchains. These two concepts are distinct and function differently. Private Blockchains control the nodes allowed to submit transactions to the Blockchain, while in permissioned Blockchains, there is regulation of the nodes allowed to verify Blockchain transactions.

Private Blockchains are private to entities who authorized within a particular group such as an organization, where only members of that group can write to or access the Blockchain. In public Blockchains, on the other hand, any entity is able to access the Blockchain and posttransactions to it. This is a form of *group-based privacy of information and party*: the group here is the organization to which the Blockchain is private.

In permissioned Blockchains, the nodes allowed to participate in the verification process are selected by a central authority, while in permissionless Blockchains no authorization is required, and any entity can participate in the verification process. In return for some monetary reward, any entity can contribute their mining computational power to the verification process. We illustrate the concept of a permissioned and permissionless Blockchain in Fig. 9. As shown, all nodes in permissionless Blockchains can access the Blockchain network.

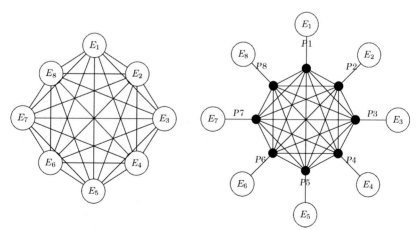

Fig. 9 Permissionless (*left*) and permissioned (*right*) Blockchains. Entities *E* need permission P to access the Blockchain.

Nodes in permissioned Blockchains can access the Blockchain network subject to some permission. This permission could be in the form of key used to access the Blockchain network, or of permission obtained by some authority for the network. This is another form of *group-based privacy of information and party*: the group who can access any information posted to, or the identities of those posting to, the Blockchain (subject to that information being further obscured) consists of all those nodes who are permissioned.

4.2.2 Encrypting Posts to the Blockchain

Generally, in implementations of Blockchain, the data in each block is stored as hashes. Merkle trees enable the storage of multiple pieces of data as a hash value calculated from the hashes of this data [1]. In a Merkle tree, each non-leaf node is the hash of the values of its child nodes. We illustrate the way in which Merkle trees work in Fig. 10.

Storing the data in each block as a hash means that it is not viewable in plain sight through retrospective inspection of the Blockchain, yet the data cannot be changed without altering the hash value (the "Top Hash" value in Fig. 10), and thus the block header, and thus all following blocks. This means the data, while hidden, remains tamperproof. This gives *total privacy of information* in storage, unless the hash is cracked through brute force.

By these means, verified data is often obscured for storage in implementations of Blockchain. There are several approaches to hiding the data at the stage when it is initially posted to the Blockchain. The most basic of these is encryption of the data. Documents can be encrypted before posting to the Blockchain using the public key of its owner, such that only the owner with their private key can see the documents. This could be suitable for the case of using Blockchain simply for the secure storage of data records:

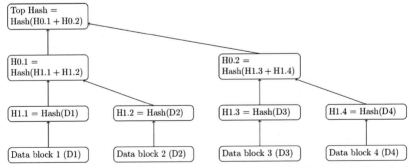

Fig. 10 Merkle tree. The data in a block (D1, D2, D3, D4) are hashed in pairs, iteratively, to produce a "Top Hash" value for the block.

an owner has control over their own data (which is stored in this ledger which is tamperproof and distributed backed up through its distribution) and can view their data using their own private key. If a user wishes to grant the permission to another party to also view their data, they can do so by sharing their key with that entity.

In cases where it is necessary that the ability to decrypt a piece of data is more distributed than the above (that a single entity cannot decrypt a piece of data) secret sharing encryption schemes can be used. Data posted to the Blockchain can be encrypted using schemes such as Shamir secret sharing [53], such that a threshold number of entities, and no fewer, must cooperate to decrypt the data. This could be appropriate for ensuring the security of sensitive pieces of data, ensuring that no entities owning keys can decrypt the data without the consent of others. This raises the barrier to accessing the data maliciously and means that the compromise of a single entity's key does not lead immediately to compromise of the data.

Cryptographically secure methods of obfuscation enable programs to run as their "black-box" equivalents. While the program functions in the same way, producing the same outputs given the same input, the details of the way the program works are not revealed by the process. This concept is a "black-box": inputs go in and outputs are generated, with no details of how these outputs were generated revealed. Perfect black-box obfuscation is mathematically impossible, yet indistinguishability obfuscation is possible and may be applicable for some cases [52]. That is, given two equivalent programs, it is impossible to tell which output came from which source. Mechanisms are known for this type of obfuscation, but are extremely inefficient.

Secure multiparty computation [54] is another approach that is as good as obfuscation, as long as there is an honest majority of participants [52]. It enables a program to be split between N parties, such that a threshold number T of these parties need to cooperate to reveal its state or complete a computation. If T constitutes a majority of the parties (is greater than $N/2$), then this is equally as good as obfuscation, given a majority of honest parties. Secret sharing and secure multiparty computation give a form of *group-based privacy of party*, in which the group must cooperate to reveal the information.

More specifically, secure multiparty computation splits data among N parties (using secret sharing), who can each carry out computations on their share of the data, resulting in the result being secret shares between the parties, without ever having been, or having to be, computed or kept on any one single device. This is an effective way of obfuscating computations and the workings of programs, since it prevents any one single party from knowing the entire computation—a verification of a transaction, for example. No single party needs to know all of the data to be verified, and thus, the computations that need to be carried out on data posted to the Blockchain can be decentralized.

Zero-knowledge proofs are another concept with use in this application. These are mathematical proofs that can be constructed to show that a particular output should result from running a given program on some input, without revealing any other information about the program. The Ethereum Blockchain platform is currently incorporating zero-knowledge proofs through the ZoKrates project, for example [55].

Succinct Arguments of Knowledge (ZK-SNARKS) allow the construction of zero-knowledge proofs for any function. Using ZK-SNARKS could enable the creation of programs in which an output is known based on conditions to be verified, for example, and outputs only that verification value, with no information revealed about the actual values that were input into the program, other than the fact that they satisfied the verification algorithm. Running such programs could be used to enable the verification of users of processes on the Blockchain: that they meet certain requirements, for example, without actually revealing any other information pertaining to the user or the values they inputted.

4.2.3 Hiding the Origin of Posts: Short-Term Keys and Pseudonyms

It may be necessary to hide not only the data posted, as discussed in Section 4.2.2, but also the entity posting it (*privacy of party*). If there is no network membership requirement or need to establish a network pseudonym (in order to retain a trail of actions by entities on that network), then this is trivial: each transaction can be made from a different one-time anonymous account. This use of one-time accounts is the approach adopted to achieving anonymity of transactions in Bitcoin, in which each transaction empties one Bitcoin account. Users generate a new private key for each new account into which they are receiving funds [1]. This means that each one of a user's funds is not linked to his other funds by default, and this approach achieves *total privacy of party*.

Using this mechanism in Bitcoin means that, in effect, each new account is a pseudonym for that user, and the pseudonyms cannot be linked. In this way, entities carrying out transactions (making payments) using Bitcoin can achieve a level of Pseudonymity in which none of their actions are linked. A problem with this is that if at any point, e.g., in the future, a transaction is made from two accounts simultaneously, then they can be linked. Techniques to mitigate this were developed: Merge Avoidance [56] and CoinJoin [57].

Some applications have a more complex requirement, in which the actions of an entity should be linked, or traceable under certain conditions (*conditional privacy of party*), without revealing their true identity. In the following, we explore approaches to obscuring the identity of parties in a Blockchain network, while maintaining some concept of network membership: proof of certain network membership conditions, retention of links between transactions, or the ability to trace links between transactions under certain conditions.

To enable a presence (an account or transaction address, for example) to be established, and actions to be linked, pseudonyms can be used. Parties can use pseudonyms that do not reveal their actual identity. In this way, the actions of an entity in a network can be linked to the same pseudonym, and transactions can be directed at their accounts, without revealing their identity. In practice, these pseudonyms can take the form of private keys possessed by an entity, for example.

There are cryptographic methods of enabling parties to prove network membership without revealing their true identity. One such approach is to use ring signatures, which enable a signer to prove that they are authorized on a network (that their private key responds to a set of public keys) without revealing exactly which public key it corresponds to [58]. This can enable entities to prove that they are authorized on networks, without revealing any more information about their identity. For networks in which membership is required, but anonymity desirable—such as anonymous forums, for example—this is a possible approach. As in the use of one-time accounts in Bitcoin, this approach can achieve *total privacy of party*, yet allows network membership verification.

Linkable ring signatures extend the concept of ring signatures, by revealing if an entity signs twice with the same private key, but again not revealing any further information [59]. This means that it can be proved that the same entity in a network has carried out two separate actions—their actions can be linked—without revealing their identity. This could enable the retention of some trail of transactions made by network members, who can prove that they are authorized, and whose actions can be linked, without ever revealing their true identity.

There are applications in which it is necessary to prevent the linking of the actions carried out by network entities, while retaining proof of network membership. This is particularly true in ad hoc networks, in which a user may carry out multiple actions across different locations, for example, and we discussed the particular case of tracking in vehicular ad hoc networks in detail in Section 3.4.1. Ubiquitous and IoT networks share this concern: a user's actions across their multiple devices on which they carry out trans- actions should not necessarily be linked—for example, the actions of a user on their laptop with a smart appliance such as a fridge.

In such applications, where preventing linking of transactions in nec- essary, linkable ring signatures are a solution. Another approach is to use short-term pseudonym updates: a user's pseudonyms can take the form of short-term keys, which update at regular time intervals. It is then nec- essary that these short-term keys cannot be linked together by their updates. Axon and Goldsmith proposed an unlinkable short-term key update mech- anism for Blockchain, applied to building PKIs in particular [40]. Here, the short-term key updates of a network entity cannot be linked publicly, but are verified by a subgroup of the network (achieving *group-based privacy of party*). Identities can be disclosed if necessary (in legal cases, for example) by the network entity in question, or a network majority, to prove the chain of actions taken by a particular party.

In existing cryptocurrencies, there have been a range of methods proposed and used for obfuscating the true identity of the originator of transactions. The use of some of these methods can extend to nonmonetary applications of Blockchain. An example is Zerocash, which uses Decentralized Anonymous Payment (DAP) schemes, developed to enable anonymous payments [60]. DAP schemes leverage ZK-SNARKs, to enable users to directly pay one another privately, with the amount transferred, and the origin and destination of the payment made, hidden in the transaction. Monetary transactions are not linked to the payment's origin: parties must prove only that they own one of the accounts, in a set, that has not yet been used, and need not reveal which one. While promising, Kosba wrote that these approaches may undermine programmability, noting a difficulty in developing approaches that are pro- grammable without exposing transactions in clear text [50].

4.3 The Trade-Off Between Security and Privacy in Blockchain Applications

We consider the balance between security and privacy: there are cases in which privacy requirements and the mechanisms needed to achieve them counter the security properties that made Blockchain advantageous in the first place. It is perhaps not suitable to use Blockchain to build systems in applications for which this is the case. We explore the trade-off between the security advantages of Blockchain we described in Section 2.1, and the privacy requirements and mechanisms highlighted in Sections 3 and 4, respectively. In Table 1, we reflect on the three case studies we explored in Sections 3.2.1, 3.3.1, and 3.4.1, summarizing for each the security advantages of using Blockchain, the types of privacy required, and possible approaches to achieving an appropriate balance.

There are clear juxtapositions between the inclusion of many of the described privacy mechanisms in applications using Blockchain, and the original security advantages of using technologies based on Blockchain. The decentralization of data and power that Blockchain-based systems can offer, which is a key advantage promoting its use in many applications such as networks and data storage, would be reduced by levels of privacy in which information was not viewable by other parties, or accessible to parties only under certain conditions.

In applications where ensuring that the privacy of the parties involved is maintained is more important than enabling auditability of ledgers by a wide array of parties (such as in the case of medical records, as we discussed in Section 3.2.1) careful consideration should be given as to the types of Blockchain appropriate (most likely a Blockchain private to the health care provider), and the parties that should have access to this Blockchain.

The transparency that is a key advantage of Blockchain for certain applications would not be retained in a system in which the information posted to the Blockchain was hidden. By in any way obfuscating the data posted, or the parties posting it, we decrease the natural benefits of observation on the Blockchain by reducing the transparency of the processes carried out on it. The auditability of the ledger Blockchain can produce is, as we discussed in Section 2.1, one of its key advantages for applications in cybersecurity. It has the potential to give users agency over the data they store on it and the ability to view its state, as well as enabling identification of misbehaving users in regulation of networks. This transparency of actions, however, is clearly a contradiction with privacy requirements.

In a number of applications, there is a need to obscure the data posted by parties, or the identity of those parties. There are approaches to doing so while maintaining a level of transparency: as we discussed in Section 4.2, conditional or group-based privacy of information or party can enable transparency toward only certain parties, or under only certain conditions. As such, a level of privacy can be maintained, while the benefits of observation for the auditability of regulation of processes are still enabled to some extent.

Table 1 A Summary of the Blockchain Security Advantages and the Privacy Requirements for Each Case Study

Use-Case	Security Advantages	Privacy Requirements	Possible Approaches
Medical records	– Auditing of the contents of medical records stored by all parties with permissions to view posts or updates. Using Blockchain, this information can be recorded reliably, and tampering with records prevented – Patient agency over their data—e.g., the potential for patients to participate in allowing access to their data to certain parties by sharing keys	– Confidentiality of the contents of patients' medical records, which are highly sensitive, is crucial – Confidentiality of the frequency and times of updates to the medical records of patients is required (i.e., avoiding leakage of information pertaining to the frequency of medical checkups)	– Given the sensitivity of medical data, it is appropriate to use Blockchains private to health care providers – Group-based privacy of party and information, in which groups include patients and doctors, and might include health care authorities or impartial network members for auditing purposes – Conditional privacy of party, in which parties are granted access to the data stored in the Blockchain in particular cases. E.g., regulators may view the identity of parties in cases such as suspected data tampering or nonconsensual use of data, and medical professionals may be granted access to a patient's medical information in life-threatening situations – Such conditions could be recognized automatically (e.g., by querying a database) or by a regulating authority for the private Blockchain, and permissions thus granted
Smart contract voting	– The transparency of a Blockchain-based voting log could enable public auditing of voting processes and verification of outcomes – The reliability of the ledger maintains the integrity of the votes made and prevents post hoc tampering with votes	– Confidentiality of a vote cast by a party: voting should be totally anonymous and a vote cast should reveal no information pertaining to its origin to any party – Metadata associated with a vote cast (e.g., time and location of voting) may reveal information as to its origin. Revealing such metadata should be avoided	– Public Blockchains, or at least Blockchains that can be accessed by all voting entities, would enable auditing of the voting process by all voting parties. This transparency is one of the key advantages of using Blockchain in this case – No privacy of information is required: the votes themselves can be cast in plain sight – Total privacy of party is needed: no information pertaining to the identity of the party who cast a vote should be revealed. It must be ensured that no party can cast a vote more than once, however. Linkable ring signatures are a mechanism that could enable this level of privacy of party, while revealing if any party votes more than once

Vehicular networks	– Decentralizing communications over vehicular networks would prevent any single party (malicious, hacked, or otherwise) from viewing all information and thus tracing the actions of parties – The ability to audit the reliable record of communications given by Blockchain is advantageous here, since reliable tracing of past actions is required in cases of misbehavior on vehicular networks	– The communications over a vehicular network by an entity should not enable the remote tracking of the location or actions of that entity – The contents of the safety messages themselves (e.g., the fact that a vehicle is braking) need not be confidential, as long as they do not reveal the identity of the communicating vehicle to the network – Authorities may need to trace vehicular communications and their origin in cases of misbehavior	– Public Blockchains are appropriate: given the frequency of the movement of vehicles, assigning vehicles to private Blockchains would be complex. Alternatively, location-based private Blockchains could be used, where on moving into a new location (e.g., city), vehicles join its private Blockchain network – No privacy of information is appropriate for safety message contents – Group-based privacy of party is suitable, in which it may be appropriate for groups to be formed based on location (e.g., using distance-bounding protocols). Groups can verify the correctness of communications from a particular entity – Conditional privacy of party is required, to enable tracing of entities is case of misbehavior in vehicular networks. Tracing Managers (e.g., policing or legal authorities) may need to access the communications of a particular party of verify misbehavior on the network (e.g., sending misinformation, or flooding the network to deny service). Disclosure of information or keys by the vehicle in question, by a network majority, or through the use of secret-sharing schemes could enable this conditional tracing

There are, however, applications where these levels of transparency may be inappropriate—in the storage of sensitive company data, for example, or the regulation of highly private governmental networks. In such cases, in which the need for privacy far outweighs any requirement for transparency, privacy mechanisms would need to be implemented (both to maintain privacy of information and of party) to such an extent that the original benefits of using a Blockchain-based system (its decentralization and transparency, and the reliability of the transaction ledger that is produced, in particular) would largely be lost. There would be little sense in using Blockchain to develop systems in applications in which this is the case.

5. CONCLUSION

We reviewed proposed applications of Blockchain-based systems for maintaining data integrity, constructing networks, and managing and executing contracts. The properties afforded by Blockchain-based systems have been shown to have a range of potential advantages in such applications. The requirements for privacy within these applications differ. In some applications, such as the storage of sensitive data, privacy is of paramount importance; in others, such as the regulation of vehicular networks, the privacy of some message content is not suitable—for safety messages, for example. The nature of the privacy needed here is more complex, conditional on location and on requirements for tracing misbehavior. We explored appropriate privacy provisions for the targeted applications and compared them with the natural privacy provision of Blockchain.

Based on the potential advantages of using Blockchain in the cybersecurity applications that have been proposed, the extent to which privacy is required in these applications, and the nature of the privacy needed, we presented approaches to building suitable Blockchain-based systems. We highlighted existing technologies by which the security and privacy requirements for these applications could be achieved in Blockchain-based systems. By considering the effect of privacy-enabling technologies on the original security advantages of using Blockchain, we found that there are certain types of application—in particular the regulation of highly private, sensitive data, or networks—in which the use of Blockchain technology may not be appropriate.

In summary, in developing any Blockchain applications in cybersecurity, it is important to balance the benefits of using the Blockchain: decentralization, transparency, and the production of reliable ledgers, with the challenges this produces for maintaining data confidentiality, or the privacy of parties using the system. This trade-off needs to be considered for each individual application of Blockchain technology, to inform its design and assess whether the security and privacy properties that Blockchain can afford are appropriate.

KEY TERMINOLOGY AND DEFINITIONS

Privacy-awareness Privacy-awareness refers to the extent to which parties have control over the disclosure of their information. This information could pertain to the contents of their communications, the actions they take, or their identity, for example.

Smart contracts Smart contracts are programs that can be run using Blockchain, which define the rules governing transactions. The rules required for a contract—the terms of an insurance contract, for example—can be mutually agreed and stored using a Blockchain ledger. The automatic execution of contractual transactions—such as the automatic payment of an insurance customer under agreed conditions—can also be executed by the programs run on the Blockchain.

Vehicular ad hoc networks In emerging Intelligent Transport Systems, vehicular ad hoc networks are used to facilitate communications between intelligent vehicles (driverless cars, for example). Safety messages, such as information about the intended actions of a car—its decision to brake or change direction, for example—are communicated between vehicles using these networks. Vehicular ad hoc networks also facilitate the communication of messages from authorities and roadside infrastructure to vehicles, such as information about travel conditions.

REFERENCES

[1] Nakamoto, S, Bitcoin: A Peer-to-Peer Electronic System, Bitcoin [online]: https://bitcoin.org/bitcoin.pdf, accessed on 10-02-2018, 2008.

[2] H. Halpin, M. Piekarska, in: Introduction to security and privacy on the Blockchain, IEEE European Symposium on Security and Privacy Workshops (EuroS&PW), IEEE, 2017, pp. 1–3.

[3] Zyskind, G., and O. Nathan, Decentralizing privacy: using Blockchain to protect personal data, In IEEE European Symposium on Security and Privacy Workshops (EuroS&PW), IEEE (2015).

[4] J.D. Halamka, A. Ekblaw, The potential for Blockchain to transform electronic health records, Harvard Business Review (2017), https://hbr.org/2017/03/the-potential-for-blockchain-to-transform-electronic-health-records (accessed 09.04.2018).

[5] O. Williams-Grut, Estonia Is Using the Technology Behind Bitcoin to Secure 1 Million Health Records, [online]: http://www.businessinsider.com/guardtime-estonian-health-records-industrial-Blockchain-bitcoin-2016-3?r=UK&IR=T, accessed on 10-02-2018, 2016.

[6] A. Ekblaw, A. Azaria, J.D. Halamka, A. Lippman, in: A Case Study for Blockchain in Healthcare:"MedRec" prototype for electronic health records and medical research data, Proceedings of IEEE Open & Big Data Conference, 2016.

[7] Ekblaw, A., Azaria, A., Vieira, T. and Lippman, A., MedRec: Medical Data Management on the Blockchain, PubPub, 2016.

[8] K. Peterson, R. Deeduvanu, P. Kanjamala, K. Boles, in: A Blockchain-based approach to health information exchange networks, Proceedings of the NIST Workshop Blockchain Healthcare, 1, 2016, pp. 1–10.

[9] Deleted in review.

[10] X. Yue, H. Wang, D. Jin, M. Li, W. Jiang, Healthcare data gateways: found healthcare intelligence on Blockchain with novel privacy risk control, J. Med. Syst. 40 (10) (2016) 218.

[11] G.W. Peters, E. Panayi, Understanding modern banking ledgers through Blockchain technologies: future of transaction processing and smart contracts on the internet of money, in: Banking Beyond Banks and Money, Springer International Publishing, 2016, pp. 239–278.

[12] K. Delmolino, M. Arnett, A. Kosba, A. Miller, E. Shi, in: Step by step towards creating a safe smart contract: lessons and insights from a cryptocurrency lab, International Conference on Financial Cryptography and Data Security, Springer Berlin Heidelberg, 2016, pp. 79–94.

[13] N. Szabo, Formalizing and securing relationships on public networks, First Monday 2 (9) (1997), http://journals.uic.edu/ojs/index.php/fm/article/view/548/469, (accessed 09.04.2018).

[14] Buterin, V., A Next-Generation Smart Contract and Decentralised Application Platform, White Paper [online]: https://github.com/ethereum/wiki/wiki/White-Paper, (accessed on 10-02-2018), 2014.

[15] F. Idelberger, G. Governatori, R. Riveret, G. Sartor, Evaluation of logic-based smart contracts for Blockchain systems, in: International Symposium on Rules and Rule Markup Languages for the Semantic Web, Springer, Cham, 2016, pp. 167–183. July.

[16] T. Nugent, D. Upton, M. Cimpoesu, Improving data transparency in clinical trials using Blockchain smart contracts, F1000Research 5 (2016), https://f1000research.com/articles/5-2541/v1, (accessed 09.04.2018).

[17] P. Sreehari, M. Nandakishore, G. Krishna, J. Jacob, V.S. Shibu, in: Smart will converting the legal testament into a smart contract, Proceedings of the International Conference on Networks & Advances in Computational Technologies (NetACT), IEEE, 2017, pp. 203–207.

[18] Authoreon.[online]: https://www.authoreon.io/, accessed on 10-02-2018, 2016.

[19] S. Higgins, Abu Dhabi Stock Exchange Launches Blockchain Voting, Coindesk [online]: http://www.coindesk.com/abu-dhabi-exchange-Blockchain-voting/, (accessed on 10-02-2018), 2016.

[20] P. Boucher, What if Blockchain technology revolutionised voting? in: Scientific Foresight Unit (STOA), European Parliamentary Research Service, 2016.

[21] P. Aradhya, Distributed Ledger Visible to All? Ready for Blockchain? Huffington Post [online]: https://www.huffingtonpost.com/pradeep-aradhya/are-we-ready-for-a-global_b_9591580.html, (accessed on 10-02-2018), 2016.

[22] A. Hertig, The First Bitcoin Voting Machine Is on Its Way, Motherboard Vice [online]: http://motherboard.vice.com/read/the-first-bitcoin-voting-machine-ison-its-way, (accessed on 10-02-2018), 2015.

[23] B. Wire, Now You Can Vote Online With a Selfie, Business Wire [online]: http://www.businesswire.com/news/home/20161017005354/en/VoteOnline-Selfie, (accessed on 10-02-2018), 2016.

[24] P. McCorry, S.F. Shahandashti, F. Hao, in: A smart contract for boardroom voting with maximum voter privacy, International Conference on Financial Cryptography and Data Security, Springer, Cham, 2017, pp. 357–375.

[25] D. Boneh, in: The decision Diffie-Hellman problem, International Algorithmic Number Theory Symposium, Springer, Berlin, Heidelberg, 1998, pp. 48–63.

[26] U. Feige, A. Fiat, A. Shamir, Zero-knowledge proofs of identity, J. Cryptol. 1 (2) (1988) 77–94.

[27] M. Coblenz, in: Obsidian: a safer Blockchain programming language, Proceedings of the 39th International Conference on Software Engineering Companion, IEEE Press, 2017, pp. 97–99.

[28] Obsidian Messenger Beta Update, [online]: https://obsidianplatform.com/2017/12/08/obsidian-messenger-beta-update/, (accessed on 08/01/2018), 2017.

[29] P. Taylor, Applying Blockchain Technology to Medicine Traceability, Securing Industry [online]: https://www.securingindustry.com/pharmaceuticals/applying-Blockchain-technology-to-medicine-traceability/s40/a2766/#.V5mxL_mLTIV, (accessed on 10-02-2018), 2016.

[30] A. Hari, T.V. Lakshman, in: The internet Blockchain: a distributed, tamper-resistant transaction framework for the internet, Proceedings of the 15th ACM Workshop on Hot Topics in Networks, ACM, 2016, pp. 204–210.

[31] C. Fromknecht, D. Velicanu, S. Yakoubov, A Decentralised Public Key Infrastructure with Identity Retention, IACR Cryptol. ePrint Archiv. (2014) 803–819.

[32] M.R. Thompson, A. Essiari, S. Mudumbai, Certificate-based authorization policy in a PKI environment, ACM Trans. Inform. Syst. Secur. 6 (4) (2013) 566–588.

[33] N. Leavitt, Internet security under attack: the undermining of digital certificates, Computer 44 (12) (2011) 17–20.

[34] A. Barenghi, A. Di Federico, G. Pelosi, S. Sanfilippo, in: Challenging the trustworthiness of PGP: is the web-of-trust tear-proof? European Symposium on Research in Computer Security, Springer, Cham, 2015, pp. 429–446.

[35] S.B. Roosa, S. Schultze, Trust darknet: control and compromise in the internet's certificate authority model, IEEE Internet Comput. 17 (3) (2013) 18–25.

[36] S. Matsumoto, R.M. Reischuk, in: IKP: turning a PKI around with decentralised automated incentives, IEEE Symposium on Security and Privacy (SP), IEEE, 2017, pp. 410–442.

[37] B. Qin, J. Huang, Q. Wang, X. Luo, B. Liang, W. Shi, Cecoin: a decentralised PKI mitigating MitM attacks, Future Generation Computer Systems (2017), https://www.sciencedirect.com/science/article/pii/S0167739X17318381, (accessed 09.04.2018).

[38] D. Wilson, G. Ateniese, in: From pretty good to great: enhancing pgp using bitcoin and the Blockchain, International Conference on Network and System Security, Springer International Publishing, 2015, pp. 368–375.

[39] P. Zimmermann, Pretty good privacy: public key encryption for the masses, in: Building in Big Brother, Springer-Verlag New York, Inc., 1995, pp. 93–107.

[40] L. Axon, M. Goldsmith, in: PB-PKI: a privacy-aware Blockchain-based PKI, Proceedings of the International Conference on Security and Cryptography, ICETE, 2017.

[41] Gartner, Gartner Says 6.4 Billion Connected 'Things' Will be in Use from in 2016, Up 30 Percent From 2015 [online]: http://www.gartner.com/newsroom/id/3165317, (accessed on 10-02-2018), 2015.

[42] A. Dorri, S.S. Kanhere, R. Jurdak, P. Gauravaram, in: Blockchain for IoT security and privacy: the case study of a smart home, IEEE International Conference on Pervasive Computing and Communications Workshops (PerCom Workshops), IEEE, 2017, pp. 618–623.

[43] M. Samaniego, R. Deters, in: Blockchain as a service for IoT, IEEE International Conference on Internet of Things (iThings) and IEEE Green Computing and Communications (GreenCom) and IEEE Cyber, Physical and Social Computing (CPSCom) and IEEE Smart Data (SmartData), IEEE, 2016, pp. 433–436.

[44] K. Christidis, M. Devetsikiotis, Blockchains and smart contracts for the internet of things, IEEE Access 4 (2016) 2292–2303.

[45] M. Conoscenti, A. Vetro, J.C. De Martin, Blockchain for the internet of things: a systematic literature review, in: IEEE International Conference on Computer Systems and Applications, IEEE, 2016.

[46] G.C. Polyzos, N. Fotiou, in: Blockchain-assisted information distribution for the internet of things, IEEE International Conference on Information Reuse and Integration (IRI), IEEE, 2017, pp. 75–78.

[47] S. Huh, S. Cho, S. Kim, in: Managing IoT devices using Blockchain platform, International Conference on Advanced Communication Technology (ICACT), IEEE, 2017, pp. 464–467.

[48] S. Rowan, M. Clear, M. Gerla, M. Huggard, C.M. Goldrick, Securing vehicle to vehicle communications using Blockchain through visible light and acoustic side-channels, arXiv Preprint arXiv 1704 (2017) 02553.

[49] P.K. Sharma, S.Y. Moon, J.H. Park, Block-VN: a distributed Blockchain based vehicular network architecture in Smart City, J. Inf. Process. Syst. 13 (1) (2017) 184–195.

[50] A. Kosba, A. Miller, E. Shi, Z. Wen, C. Papamanthou, Hawk: the Blockchain model of cryptography and privacy-preserving smart contracts, in: In IEEE Symposium on Security and Privacy (SP), IEEE, 2016, pp. 839–858.

[51] ETSI, Intelligent Transport Systems (ITS); Security; Threat, Vulnerability and Risk Analysis (TVRA), Etsitr 102 893 v1.1.1 (2010−03), Technical Report, 2010.

[52] Buterin, V., Privacy on the Blockchain, Ethereum Blog [online]: https://blog.ethereum.org/2016/01/15/privacy-on-the-Blockchain/, (accessed 10-02-2018), 2016.

[53] A. Shamir, How to share a secret, Commun. ACM 22 (11) (1979) 612–613.

[54] A.C. Yao, in: Protocols for secure computations, Annual Symposium on Foundations of Computer Science, SFCS'08, IEEE, 1982, pp. 160–164.

[55] Eberhardt, J, ZoKrates—A Toolbox for zkSNARKs on Ethereum, Ethereum Foundation [online]: https://ethereumfoundation.org/devcon3/sessions/verifiable-off-chain-computation-for-smart-contracts/, (accessed on 10-02-2018), 2017.

[56] Hearn, N., Merge Avoidance: Privacy Enhancing Techniques in the Bitcoin Protocol, Coindesk [online]: http://www.coindesk.com/merge-avoidance-privacy-bitcoin/, (accessed on 10-02-2018), 2013.

[57] Maxwell, G., CoinJoin: Bitcoin Privacy for the Real World, Bitcoin Forum [online]: https://bitcointalk.org/index.php?topic=279249.0, (accessed on 10-02-2018), 2013.

[58] R.L. Rivest, A. Shamir, Y. Tauman, in: How to leak a secret, International Conference on the Theory and Application of Cryptology and Information Security, Springer, Berlin, Heidelberg, 2001, pp. 552–565.

[59] J.K. Liu, D.S. Wong, in: Linkable ring signatures: security models and new schemes, International Conference on Computational Science and Its Applications, Springer, Berlin, Heidelberg, 2005, pp. 614–623.

[60] E.B. Sasson, A. Chiesa, C. Garman, M. Green, I. Miers, E. Tromer, M. Virza, in: Zerocash: decentralised anonymous payments from bitcoin, IEEE Symposium on Security and Privacy (SP), IEEE, 2014, pp. 459–474.

ABOUT THE AUTHORS

Louise Axon is a DPhil student in Cybersecurity at the University of Oxford. She received a BA degree in Mathematics and Music from Cardiff University in 2013, and an MSc in Mathematics of Cryptography and Communications from Royal Holloway, University of London in 2014. Her research interests include approaches to monitoring network security, and the application of Blockchain technologies to security, and she has worked on the use of Blockchain for constructing privacy-aware public-key infrastructures.

Michael Goldsmith is Director of the Global Cyber Security Capacity Centre at the Oxford Martin School and a Senior Research Fellow in the Department of Computer Science and at Worcester College, University of Oxford. His research spans a wide range of topics within security, from the mathematical to the social. He received his DPhil in Computation from Oxford University three decades ago for work on support for specification logics and has also worked in concurrency theory and formal verification through exhaustive state exploration.

Sadie Creese is Professor of Cybersecurity in the Department of Computer Science at the University of Oxford. She was Founding Director of the Global Cyber Security Capacity Centre at the Oxford Martin School and a member of the Coordinating Committee for the Cyber Security Oxford network. She is engaged in a broad portfolio of cybersecurity research spanning situational awareness, visual analytics, risk propagation and communication, threat modeling and detection, network defence, dependability and resilience, and privacy.

Printed in the United States
By Bookmasters